Blacks
and
Technology
Volume II

Blacks and Technology
Volume II

The Shift of Economy Power to Blacks
in the Twenty-first Century

Technology is the jet engine of the 21st century
economy, without which Blacks will continue to
experience major economic Problems

Raymond L. Chukwu

To order additional copies of this book, contact:
Xlibris Corporation
1-888-795-4274
www.Xlibris.com
Orders@Xlibris.com
50921

CONTENTS

DEDICATION

To my grandmother, Nnn-Ihe Nwaigbo, because without her loving care, this book, among other major accomplishments in my life to this point, could not have taken place.

Madam Nnn-Ihe Nwaigbo

Also to my foster parents, Dr. William Koster and Mrs. Marcia K. Koster, because without their help to bring me back to the United States and train me as an aerospace engineer, I would not have had this opportunity to write this book

Dr William Koster and Mrs. Marcia K. Koster

The ideology that all things are subject to change is an accurate statement observed for decades. It is, therefore, logical to believe the same is applicable to the shifting of economic power to Blacks in this twenty-first century.

Acknowledgements

I want to thank the following people who contributed extensively to my present economic status:

Tennyson O. Chukwu (Nigeria)
Rev. Duggon (France)
Rev. Scott Campbell (France)
Dr. William Koster (United States)
Late Mrs. Koster (United States)
Mr. Joe King (France)
Mr. Joe Sow (Senegal)
Augstine Asiakw (France)
Francis Ironkwe (Senegal)
Mazi Nwaeyi (Cameroon)
Darlington Chukwu (Nigeria)
Uche Okoro (Nigeria)

Finally, my special thanks to my wife, Debbie Chukwu, whose support, encouragement, and motivation influenced my decision to write this book.

FOREWORD

This book, Blacks and Technology: The Shift of Economic Power to Blacks in the Twenty-first Century is very intriguing because it is the first book written to seriously disclose the reason behind the poor quality medical treatments and limited scientific and technological skills among Blacks all over the world, particularly here in the United States.

This book establishes historical facts that explain why blacks disproportionately suffer from poor health and inadequate professional health care treatment in the United States and throughout the world.

The book further provides critical insight into the reasons Blacks suffer from substandard health and receive inadequate health care treatment in America and throughout the world.

Raymond Chukwu, republic nominee for United States Congress, a finalist competitor for United States ambassador to Nigeria, an aerospace engineer, and president of Black Technologies Advancement, delivers hard-hitting evidence that, among other things, unveils the genesis of poor health care dating back to colonialism:

Why Blacks today continue to experience inadequate doctor-patient care services.

Why many false presumptions continue to exist today that connect certain serious diseases more with Blacks than other group of people.

Why leading White medical researchers continue to purposefully exclude the benefits of Africa'sa traditional healing remedies and treatments—even though such treatments have been found effective in combating the onslaught of many diseases.

The reason Blacks are supposedly faced with higher rates of illnesses, and what actions can be taken to reverse these detrimental trends.

The relationship between Africa and major drug corporations' ongoing exploitation of Africa's resources to their own economic benefits.

The economic readiness of Black against the new millennium.

Early science and technology discoveries by Blacks.

The book also examines Africa's historical role in providing other countries with medical remedies and extracts from plants and other sources still in use today.

This book should be applauded because with the exclusion of Blacks from early medicinal research, no one can give, with any certainty, the medical history of Blacks.

All the medicinal procedures and medical technologies used today are all based upon the techniques and technologies resulting from such early medicinal and medical research and developments of which Blacks were excluded.

This book comes at the right time because up to this point, no one has been able to challenge the reliability of all the medicinal and medical deficiencies attributed to Blacks, especially here in the United States.

Rather than trying to question or challenge the reliability of such disturbing medicinal trend, Blacks accept with no objection such deficiencies or hypotheses. Thus there should be high level of interest in books that could address issues of this nature. Everyone, regardless of ethnicity, should be anxious to find out the reason behind the challenging medical uncertainty, which Blacks all over the world face today.

Such medicinal or medical uncertainty has, to this point, made it very difficult if not impossible for any doctor or physician, regardless of qualifications or experience, to diagnose any Black person and accurately prescribe with any certainty the most appropriate medication or treatment without trial and error. Overall, I strongly believe that this will be a very important book due to its position in this subject area.

I have never seen anyone very concerned about the quality of Blacks' medical care and want to do something about it to help the general public. Yes, it is different, but that in itself is very refreshing.

Debbie Chukwu

PREFACE

Blacks enjoyed good health and superior medical care, with their traditional medicine and obscure natural products, before colonization.

These systems of medicine can, by no means, be considered empirical. They were based on a considerable amount of knowledge accumulated, by and large, through application of scientific method, individual observations and confirmation of these observations by others, formulation of hypotheses followed by testing of the hypotheses by experimentation. Granted that some of the tools employed were primitive but the approach was undoubtedly scientific. Accordingly, native Black Africans were able to develop and introduce remarkable new drugs for treatment against various diseases, from plants and special-illness-healing herbs, long before colonization.

That is to say that native Black Africans were already well equipped to confront and challenge any medical problems among its people for the past several centuries. Such capabilities led to miraculous successes in the control and treatment of many diseases in sub-Saharan Africa. These efforts contributed significantly to the continued popularity of plant-and-herb remedies among native Black Africans, in part, because of the ready availability of these plants or herbs.

Native sub-Saharan African herbs and plants achieved miraculous successes against various diseases long before the colonization.

However, colonization seriously disrupted these miraculous successes and changed completely the dynamics of the overall equation of unconventional and traditional treatments with the native Black African herbs or plants.

Considering the successes of these remedies, it was absolutely no way for the sub-Saharan Africans to give up such effective remedies without a fight.

I cannot emphasize enough the urgency for Blacks to engage themselves with science- or technology-related undertakings. Native Black Africans' limited access to technology is the root cause of Blacks problems all over the world today. I will tell you in a second why science or technology is the root cause of the current humiliation experienced by Blacks all over the world today.

Before colonization, native Black Africans' most sophisticated defense weaponries were bows and arrows. These bows and arrows were sufficient for hunting purposes.

Realizing the abundant riches and the peacefulness of the Africans, the colonizers had no choice but to utilize the superior capabilities of their weaponries to launch an all-out assault on the sub-Saharan Africans to overtake their land. The assault compounded by the inferior defense capabilities with bows and arrows marked the beginning of the fall of the Black civilization and the rise of the White civilization.

As you will see in this book, the fall of Blacks and the rise of Whites will be the principal reason for the sub-Saharan Africa to be medically isolated from the early scientific research and development.

In the sixties, two prominent Black leaders emerged to liberate the Blacks from their suppression of the Whites. Dr. Nnamdi Azikiwe, according to his autobiography, met secretly with Dr. Martin Luther King, Jr., to discuss the negative influence of the White suppression. Dr. Azikiwe, with an advanced degree in political science from the United States, returned to Nigeria in 1959 to lead the effort to expel Whites from sub-Saharan African countries.

Dr. Martin Luther King, Jr., watched this movement with keen interest, and luckily, the effort was well executed. In 1960, Nigeria, the biggest and richest country in Black Africa, received her independence from her British master. Dr. Azikiwe became the first Nigerian president.

His method of approach inspired Dr. Martin Luther King Jr., to lead Black Americans against racial injustice and White suppression in the United States in 1963/64. This was partly successful, because racial justice was achieved, but the suppression continued.

Sadly, Dr. Azikiwe has remained obscured in history, and certainly, I have no explanation for that. The important thing to note is that in these two instances, Blacks were successful because they were united as a team and fought for a defined objective. In the history of Blacks, this was probably the only time Blacks were united as a team to achieve a definite purpose.

Apart from Dr. Azikiwe and Dr. King, no Black leader has emerged since then to unite the Blacks as a team to fight against the current economic poverty experienced by Blacks all over the world. The achievements of Blacks, since then, have been on an individual basis, which has not contributed to minimizing the economic struggle among Blacks. Since then, no Black leader has emerged to lead this new crusade. With technological innovations, it will be my dedicated life mission to undertake such task at this moment when Blacks are at crossroads. Before the end of this century, the Blacks would have accepted the bad experience of slavery as a challenge to greater effort needed to generate the necessary momentum needed to unite the Blacks as a team and fight against present poverty through the utilization of sub-Saharan African untapped, obscured natural and mineral resources.

Raymond Chukwu

THE POEM OF THE STRUGGLING BOY FROM SUB-SAHARAN AFRICA

My God, my God
Of all the Races, You Created
Black race is the coolest one,
Because of abundant natural resources
My God, My God,
Of all the races I know,
Black race Worship you the most
Yet Blacks face the worst poverty
Why, Why, My God, my God,
My God, My God,
Of all the continents I know,
The wealthiest is Black Africa
Because of its natural resources
My God, My God,
Of these blessings and Worshiping
Why, Why, Why, My God,
Do Black race suffer the worst poverty
My God, My God, My God
Why, Why, Why, Why.

INTRODUCTION

Is technology against Blacks, or are Blacks against technology? Either way, Blacks must be identified with science- or technology-related products before they can overcome the current frustrating and shameful economic poverty, facing a large percentage of their population. The poverty of Blacks can be attributed to the fact that Blacks as a race are not associated with any science- or technology-related developments, which is the prerequisite for economic security, without which any race will experience difficult economic growth.

Before developing any type of solution to any kind of a problem, the cause of the problem must first be identified. Over the decades, Blacks, including their leaders, have fought for equalities, but failed significantly to achieve economic equality because they have been unable to correctly identify the cause of their economic problems.

The abundant agricultural, natural, and minerals resources of sub-Saharan African nations make sub-Saharan African nations a segment of significant importance in the twenty-first-century world economy, given the rapid depletion of the world resources at this point in time. Effective utilization of these resources to build stronger social and economic relationship between sub-Saharan Africa and the United States, with Nigeria as the model, should be an important social and economic issue for the United States and other parts of the world at this moment in our history.

This book is called Blacks and Technology: The Shift of Economic Power to Blacks in the Twenty-first Century because despite decades of efforts, sub-Saharan African nations remain impoverished. Sub-Saharan African nations are poorer today than they were in 1960, sometimes by very wider margins. As a result, the sub-Saharan African nations have been the site of large-scale experiments to reform its economy.

However ambitious, these projects have failed to generate sustained economic growth. Sub-Saharan African nations—including Nigeria with its abundant natural resources, including crude oil and an inundation of a highly skilled labor force—are considered the most prosperous segment in the world economy; yet sub-Saharan Africans are as starved and malnourished while over ten million Nigerians alone with college degree in the United States, Europe, Asia, Australia, and South Africa are on self-imposed exile because Nigerian economy can not accommodate their acquired academic skills.

The self-imposed exile is frustrating to these Nigerians as over 90 percent of them find themselves with odd jobs, which pay 50 percent below their acquired academic skills. Why should over 137 million Nigerians as well as other growing population of the sub-Sahara African nations be subjected to this type of economy torture? Put differently, the sheer magnitude of this problem, which is but overwhelming, calls for an urgent and all-out intervention by an innovative approach to reform both social and economy climate in sub-Sahara Africa.

Even though progress has been made by Blacks over the years, Blacks must still achieve economic power, domestic and international prestige, and superior business creditability and advanced academic status in science and engineering before they can enjoy the same economic opportunities as their Asian and European counterparts in the United States and other parts of the world.

Asian and European Americans command and control the economic power of this nation, and that of the world, because of their commitment and interest on research and development of technology-related products. This commitment has afforded these ethnic groups with credible technology base with which to be identified here in the United States. Also, they extend the benefits of these bases to their home countries, unlike Blacks whose economic poverty has worsened over the years.

Raymond Chukwu.

CHAPTER 1

Present Economic Status of Blacks

It appears that the basic problem confronting Blacks and their leaders in this country today is their inability to realistically define the root cause of Blacks' economic problems and objectively develop realistic plans to address the problems in a more satisfactory manner. Blacks have no market share or control any type of influence either in the medicinal or medical related domains, nor other technology sectors. This is a disturbing trend and needs an honest answer. The right answer will be, Blacks do not have any market share or any influence in the world because Blacks have no products to offer. Market is a two-way street; you buy my product and I buy yours. It is as simple as that.

After reviewing the data from this book, one will realize that Blacks still have a long way to go in gaining any type of influence among the medicinal or the scientific community. In my opinion, this is not only very disappointing but also troubling. In part because Blacks, including their leaders, still do not understand that without science- or technology-related undertakings, Blacks will never be a part of the medicinal or scientific community or participate and enjoy the current economic boom in the United States.

It has been emphasized by Black leaders that it is imperative to build bridges from Wall Street to Appalachia and the Delta or prosperous Silicon Valley to the ghettos, barrios, and lonesome hillsides in the Ozarks and Middle America.

According to these Black leaders, the bridges will be built by focusing on

- ✓ access to capital
- ✓ quality education for all of our children
- ✓ the rise of jail-industrial complex
- ✓ human rights around the world
- ✓ comprehensive health care for all Americas
- ✓ inspiring voters to build coalition and shared economic interest

The only thing relative to Blacks' economic struggle in this list is the quality education. Yet no mention was made that education alone at this time is not sufficient, particularly for Blacks, if they intend to have any say in the medicinal or scientific community or gain any market share in the marketplace.

Blacks at this time must be encouraged to seek education in science, medicine, engineering, mathematics, or computer technology.

It is only with that type of educational efforts that Blacks can gain equal partnership with both the medical and scientific community or share market advantage with other ethnicities.

If Blacks have no products for sale, they must continue to buy from others with no one buying anything from them. It is therefore difficult, if not impossible, to get access to capital without a market base or formed profitable trading partners.

Any product with measurable impact at the marketplace must demonstrate scientific merit and technological feasibility. With limited participation of Blacks in science- or technology-related undertakings, Blacks cannot be identified with any science- or technology-related products; as a result, they will continue to have no market share in the world economy.

This poses the question, Is technology against Blacks, or are Blacks against technology? Either way, Blacks must first be identified with science- or technology-related products before they can successfully gain any market share.

Can Blacks enjoy superior economic status in the twenty-first century after being suppressed and oppressed for over one hundred years by the Whites? Before the slave trade and colonization of Black Africa, Blacks had superior prestige, integrity, and economic power. The slave trade and the colonization marked a miserable turning point to the superior-living standard of Blacks.

As we begin the twenty-first century, the efforts of Blacks to regain, their integrity, prestige, and superior economic status is disrupted by limited access to science and modern technology. Before colonization, Blacks had untapped natural and mineral resources. Most of these resources in Black Africa are still untapped.

This is in part due to the limited knowledge in science or technology, which is the prerequisite for economic security. Blacks lack access and knowledge to science and technology because they have not been exposed to the required type of education and training necessary to gain scientific and technological skills and knowledge that is the foundation for developing science- and technology-related products.

Table 1 has been compiled from the United States Housing and Household Economic Statistics Division, Department of Commerce to show the percentage of Black population involved or associated with engineering, science, and medical profession. According to this table, the population of this country is about 267 million, of which White population is about 220 million (82 percent) and Black population is 34 million (13 percent). Based upon these figures, there are about 1.67 million engineers in the United States, of which 1.48 million (88 percent) are White while only about 58,000 (3 percent) are Blacks. There are about 409,000 Scientists in the United States, of which about 358,000 (88 percent) are White while only about 17,000 (4 percent) are Blacks. Also there are about 875,000 doctors in the United States, of which 759,000 (87 percent) are Whites while only 28,000 (3 percent) are Blacks. Table 1 is the breakdown of these figures.

	General	White	Black	White%	Black%
Population	267 Million	220 Million	34 Million	82%	13%
Engineers	1.7 Million	1.5 Million	58,000	88%	3%
Scientists	409,000	358,000	17,000	88%	4%
Doctors	875,000	759,000	28,000	87%	3%

Table 1: Identification of the Cause of the Economic Problems of Blacks by Detailed Occupational Analysis between the Whites and Blacks

These figures should be offensive to Blacks and present a call for collective action to improve efforts to change this humiliating statistical data. The purpose of this statistical data is to explain in a very elementary way why Blacks are where they are today, in both the United States and other parts of the world economy. This data is the principal reason why Blacks have limited access and knowledge to exploring the economic advantages of their untapped natural and mineral resources.

The relationship between Blacks, drugs, and gangs can be further analyzed to support table 1. If Blacks have been deprived of science and technology, which is the prerequisite for economic power, then some Blacks might consider drugs and gangs as an alternative to enjoy the same economic power as those exposed to science and technology. Table 2 evaluates the statistical data of racial makeup of the United States population in prison. Based upon table 2, about 1.6 million of the population is in prison, of which 31 percent are Blacks.

If you consider that Blacks are only about 13 percent of the total population, then this percentage of the Black population in prison is indeed very troubling; hence, the need for science and technology for Blacks intensifies.

	General	White	Black	White%	Black%:
Population	267 Million	220 Million	34 Million	82%	13%
Imprisoned	1.6 Million	464,000	502,000	28%	31%

Table 2: Racial Makeup of the United States Population in Prison

In order to correct this unpleasant situation once and for all, the principal cause of this problem must first be identified. In this specific case, without trying to improve the statistical data in table 1, the chances to accurately resolve this complex problem facing Blacks in the United States and other parts of the world would be very difficult, if not impossible, to confront.

Blacks have a pivotal role in several of the United States major sports events but have been unable to employ this success to improve their economic status. Again this is associated with the table 1. Let me explain.

Tracks

Today, Blacks are the fastest in tracks in the world, yet they generally cannot calculate the velocity or the acceleration of their activities in a specific event. However, it is much more difficult to do the dash than to do the calculation. To illustrate, our air is composed of nitrogen, oxygen, carbon dioxide, and several atmospheric particles. This composition of the air forms the critical equation of our surrounding environment. These particles of air are constantly in motion and create opposing resistance to everyone of our movements.

Now to do the 100-meter dash, the sprinter has to overcome strong air resistance in a record time while carrying his or her own weight and maintaining total control, both in external and internal conditions. To do so requires painstaking efforts which include the following:

- ✓ vigorous training
- ✓ concentration
- ✓ focus
- ✓ self-confidence
- ✓ determination
- ✓ courage
- ✓ willpower
- ✓ self-discipline
- ✓ body dynamics (maneuverability to overcome the motion of the opposing air resistance in a record time)

On the other hand, the calculation of velocity or the acceleration of the sprinter requires neither half of the wisdom nor the skills needed to do the 100-meter dash in a record time. However, Blacks have been told that it is much more difficult to calculate the velocity or the acceleration of the sprinter, than to do the sprint itself.

If to become either an engineer or scientist took only half of the wisdom, painstaking effort, willpower, and self-discipline of what it takes to become a sprinter, believe me, we probably would have no engineers or scientists in the world today. In spite of this, Blacks are going around in life feeling inferior rather than feeling superior even though they have superior given talents and intelligence.

Sports

Let's examine the following sports dominated by Blacks in the United States.

Football:

Football is the most competitive sport in the United States, and it is dominated almost entirely by Blacks. Each team is made up of over 88 percent (88%) of Black players.

To play in this competitive sport requires the following:

- ✓ special skills
- ✓ endurance
- ✓ outsmarting your opponent
- ✓ extensive training
- ✓ first-class determination/willpower

✓ solid and strong self-confidence
✓ body dynamics (maneuverability to overcome the massive collisions associated with the game while maintaining total control of both external and internal condition necessary to win the game)

Again, it takes a lot more to become a football player than to become an engineer or scientist, but Blacks have been told otherwise.

Basketball:

Basketball is another competitive sport in the United States, and it is also dominated almost entirely by Blacks. Each team is again made up of over 88 percent (88%) of Black players.

To play in this competitive sport again requires the following similar skills as the football:

✓ special skills
✓ endurance
✓ outsmarting your opponent
✓ vigorous training
✓ self-discipline
✓ self-determination
✓ self-confidence
✓ body dynamics (maneuverability to overcome the massive collisions associated with the game while maintaining total control of both external and internal condition necessary to make the basket and win the game)

Again, it takes a lot more to become a basketball player than to become an engineer or scientist, but again, Blacks have been told otherwise.

Baseball:

Baseball is called the all-American sport. It is a very competitive sport in the United States, and is dominated almost entirely by Blacks as well. Each team is made up of over 70 percent (70%) of Black players. To play in this competitive sport requires the following skills:

- ✓ special skills
- ✓ endurance
- ✓ outsmarting your opponent
- ✓ extensive training
- ✓ first-class determination
- ✓ solid self-confidence
- ✓ body dynamics (maneuverability to pitch the ball in the most skillful manner and the concentration necessary to hit the ball with forceful impact needed)

Again, it takes a lot more to become a baseball player than to become an engineer or scientist, but the Black man has accepted fiction rather than fact.

Boxing:

Boxing is another intriguing sport mostly dominated by Blacks in the United States. Boxing is a one-on-one sport. It requires quickness and physical fitness. It also requires the necessary endurance needed to throw and take heavy punches. Boxing is almost entirely dominated by Blacks. Over 90 percent (90%) of the boxers in the United States are Black. An excellent boxer must exhibit the following skills:

- ✓ quickness
- ✓ endurance
- ✓ outsmarting your opponent
- ✓ vigorous training
- ✓ self-discipline
- ✓ self-determination
- ✓ self-confidence
- ✓ body dynamics (maneuverability to outsmart and confuse your opponent in the best manner necessary to win the fight)

	White%	Black%
Tracks	20	80
Football	12	88
Basketball	11	89
Baseball	30	70
Boxing	10	90

Table 3: Statistical breakdown of the racial composition of the United States league teams, tracks, and boxing

If you compare table 1 and table 3, you will notice an amazing similarity on the White and Black percentage in each table.

With only 13 percent of the population, Blacks dominated the field events with average percentage of 83.4 percent. This is a commendable and outstanding record. Yet with this impressive record, no significant progress has been made in this area to better the image, prestige, or the economic status of Blacks. It is sad that none of these teams is owned by Blacks.

Whites, on other hand, with 82 percent of the population, dominated science and technology with about 88 percent of engineers, 88 percent of scientists and 87 percent of doctors. This statistical data gives Whites superior economic status and the know-how to outsmart Blacks, even in sports and tracks where they dominate.

You do not need to be a rocket scientist to figure from tables 1 and 3 that technology or science is the master key to world power, superior economic status, international prestige, and improved academic status. Again, can Blacks ever restore the prestige and integrity they lost due to colonization and slave trade? The correct answer to this question is YES—with capital letters.

If it is indeed true that every adversity carries with it the seed of equivalent benefit, and if that adversity is accepted as a challenge to greater effort, then over one hundred years of devastating struggle and suffering among Blacks will definitely be rewarded in the early twenty-first century, but only if Blacks accept their suppression and oppression as a challenge to greater effort. How this will be successfully accomplished is another complex question to answer. Many answers can apply, but the most correct answer to such a complex question is the purpose of this book.

This book will describe step-by-step strategies on how Blacks can shift the economic power to their direction. The methodologies, implementation techniques are well documented in this book. It is also important to note that this is the first book ever written to tackle the poor economic status of Blacks from the science and technology viewpoint.

In order to do this successfully, the cause of the Black economic problems as demonstrated by figures in table 1 must first be identified. The figures in table 1 is a powerful tool to convincingly prove that science and technology are the prerequisites for economic security without which any race will suffer shameful economic problems. Until Blacks are identified with science- and technology-related products, their economic problems will continue.

CHAPTER 2

Relationship of Blacks with Technology and Poverty

The poverty of Blacks can be attributed to the fact that Blacks as a race do not have a technology base nor are Blacks associated or involved with any scientific or technological developments. As a result, Blacks are not identified with any science- or technology-related products and therefore cannot enjoy any of the following benefits:

- ✓ economic power
- ✓ superior defense capabilities
- ✓ domestic and international prestige
- ✓ superior business creditability
- ✓ advanced academic status in science and engineering
- ✓ advanced academic status in computer and technology

As we stated before, Blacks are unlike other races, such as the Asian and European Americans, who enjoy all of the above because they have technology bases here in the United States and consequently extend the benefits of these technological advances to their home countries, thereby commending superior economic power in the global economy.

Why are Blacks where they are today? Several answers might come to mind, but the correct answer is that Blacks are not identified with any science- or technology-related products. A second question might be why are Blacks not identified with any science- or technology-related products? The answer to the second question can be the answer to the first question as well. Every race in this world is here for a specific purpose. What is the purpose of the Black race?

Also, every race in this world has a strong and weak point. Take the White race for example on being always very ambitious with a defined mission in mind to accomplish the intended mission.

The Black race, on the other hand, with skillful talents have refused to apply their God-given talents to advance the overall economic status of Blacks all over the world instead of always thinking in terms of individual success.

Large percentage of Blacks lack the degree of patience and painstaking efforts needed to be involved with long-time, high-risk undertakings needed in the development or design of science- or technology-related products, which can result to the identification of Blacks with science- and technology-related products. Table 1 can be used to support this argument by reviewing the percentage of Blacks associated with science- and technology-related profession against that of those associated with entertainment and sports-related undertakings.

An independent study conducted by Black Technologies Advancement (BTA) revealed that about 0 percent of Black hi-tech businesses in Silicon Valley, California, were neither the prime contractors nor subcontractors to over 500 Department of Defense, and other federal government prime contractors, and approximately 3,500 subcontractors that operate in Santa Clara County, Bay Area, Northern California.

This is because a considerable gap exists between the skills acquired by White hi-tech companies and that acquired by Black hi-tech businesses to compete for opportunities emanating from the federal government, the state government, industrials, and private corporations due to limited resources, see table 1.

While the business world is a constantly changing dynamic system and no longer depends on a large stable workforce with a limited knowledge base, the success of these Black hi-tech businesses will be very challenging because it will be more difficult for them to hire or train their employees with adequate technical, manufacturing, or engineering skills needed to increase productivity and maximize profit.

Training with limited resources has failed to produce the skills needed, especially in science- and technology-related areas, because science and technology training, even those with advanced degrees, still need additional practical experience on product design and manufacturing applications, which certainly require more sophisticated facilities and resources, which in most cases exceed the budgets of the Black hi-tech businesses.

In an era of scarce resources, rapid change, and fierce global competition, Blacks will need help and assistance in order to compete successfully. Because Blacks are not well equipped with adequate financial backing, they lack sophisticated facilities and resources needed to overcome the high-cost, high-risk, and complex challenging research involved with operating in a scientific and technological environment.

The development of science/technology-related products require extensive research, which can take several years of effort, before the final product goes into the marketplace. This is due to the high risks associated with such efforts. Also, the development of the technology needed to establish the long-term national vision of an interoperable, highly complex, and dynamic product development process is too costly and too risky.

This is particularly true for Blacks who lack the necessary resources and capital to undertake such mission. The technical resources needed to tackle the problem are diverse. Blacks experience difficulties in all their business endeavors to generate funds needed to operate the business and train experts. For these businesses to acquire the requisite skills needed to compete and win competitive government grants, contracts, and subcontracts, they need some kind of government grants or funding.

Unfortunately, the government-grant laws and regulations make it very difficult, if not impossible, for any of the Black hi-tech businesses to receive the necessary research grants needed, not only to be very competitive, but to expand the business as well.

The imposed government laws and regulations on the distribution of its research grants can be seen from table 4 compiled by Black Technologies Advancement (BTA):

	White%	Black%	Hisp%	Asian%
Research grants	90	.4	1.5	8

Table 4: Federal government research grants distributions

This method of distribution of government research grants makes it very unlikely that Black hi-tech businesses will receive enough research grants needed to operate their businesses, which could make significant contribution to the economic status of Blacks through employment opportunities for the Black population.

Without an impressive employment records and massive product design and development into the marketplace, Blacks as well as Black hi-tech businesses will not make the necessary economic contributions to the Black communities. If, on the other hand, Black hi-tech businesses can create massive product design and development into the marketplace, an economic return can be achieved through the sustainable application of the technology developed while educating Black communities as to how this research and technology could be utilized at the local level to promote grassroots development and improve the general quality of life.

These Black businesses and their employees may not be at the top of economic ladder, but their earnings will certainly eliminate them from poor economic status and enable them to help other Blacks as well, thereby establishing economic security among Blacks and other ethnic minorities.

If science and technology can ever be used to address the economic problems of Blacks, the untapped Black African resources can be explored as the first step toward identifying the Black race with science- or technology-related products. If indeed we are concerned with the Black race's economic progress, we must also understand that any significant progress by the Black race must originate from sub-Saharan Africa if it is expected to make any outstanding economic contribution to the Black race.

Because most of the Black African agricultural resources, natural resources, and mineral resources are still untapped, they have remained obscure to the world market. These, untapped Black African resources have the potential to generate over $938 billion a year.

CHAPTER 3

Science and Technology of Sub-Saharan Africa

Due to the early exclusion of Black population and Black natural products from the development of any medicine or medical related undertakings, native Black African herbs or plants have not been endorsed by any medical report or literature today.

The biochemical agents of food plants and herbs from the American, European, and Asian market continue to dominate the world food supply. They also continue to be of particular interest as a source for new pharmaceuticals as well as organic products to the world community. In parallel with this analysis, the biochemical agents of food plants and herbs from the sub-Saharan African nations are not only ignored, but their traditional medicinal plants and herbs have remained obscure to the world community despite centuries of applications as effective treatment against various illnesses.

Noninclusion or the exclusion of Blacks and their natural products from the early scientific medicinal research is very regrettable.

The inclusion of effective agents from the sub-Saharan African plants or herbs would have contributed significantly to the pharmacologically applications of drugs development.

It would have also promoted the pharmaceuticals from sub-Saharan African herbs or plants as biotechnology cornerstone that can provide the following scientific and economic benefits to the United States and sub-Saharan African countries:

- ✓ biotechnology-related investment opportunities
- ✓ profitable business relationship
- ✓ wide utilization of these herbs or plants as an alternative medicine
- ✓ new scientific data to advance medical knowledge
- ✓ identification of Blacks with science-related products
- ✓ identification of Blacks with technology-related products

The following specific points can be emphasized to support the supposition that the isolation of Blacks and their natural products from medicine is a medicinal and medical catastrophe:

- ✓ Tropical plants are an important source of bioactive natural products.
- ✓ Many chemotherapeutic agents used in modern medicine are from plants or are agents derived from SAR of plant natural products

✓ Higher percentage of promising drugs isolated from herbs can illustrate the chemical diversity available in native Black African herbs or plants

✓ Black Africans' tropical forest ecosystems are threatened by development and modernization as is botanical knowledge associated with traditional cultural practices.

✓ Many of today's pharmaceuticals have their origins in plant chemistry.

These are valid points and should provide significant justification for a focused investigation with native Black African herbs and plants. This is needed to generate new concepts in order to accelerate the development of new pharmaceuticals from sub-Saharan African herbs or plants. Such investigation will proceed with a defined scientific plan aimed toward exploring the medicinal merits and market potentials of these herbs or plants in the world market.

Given the anticipated market penetration and the commercialization potentials of these sub-Saharan African herbs or plants, there is no reason whatsoever for Blacks to be where they are today. This is particularly truer with regard to their sad experience with modern medicine.

Drugs, from whatever source they may be derived, are a means to relieve suffering and to achieve and maintain good and quality health. In the past six decades the development and introduction of new drugs from plants has indeed been remarkable. Imagine the miraculous successes to have been achieved in the control and treatment of many diseases through the proper utilization and application of native Black African special-illness-healing herbs or plants in the modern medicine.

Also, the development of compounds from plant sources used in herbal or plant medicine has potential advantages in that toxicities may already be known. Moreover, if the plant or herb sources are abundant as in the case with native Black African herbs or plants and the compounds are easily isolated, costly synthesis of candidate drugs could be eliminated.

Since modern medicine came to Black Africa in the wake of colonialism, it was considered as an alien system imposed on the people by force.

With independence, there has been a renewed interest in the traditional systems of healing because it is a part of the national resurgence. The native Black African herbs or plants remedies cannot be ignored if Blacks are making the necessary efforts to take advantage of the superior pharmaceutical properties of their herbs or plants to gain market share in the new millennium.

CHAPTER 4

Potentials of Sub-Saharan African Resources

The untapped Black African agricultural, mineral, and natural resources could explode into the largest world market at the early stage of the twenty-first century. In the literature of every ancient culture, there are accounts of special agricultural, mineral, and natural resources with enormous commercialization potentialities. Unfortunately, for these other cultures, their resources are being produced and consumed at a tremendous and ever-increasing rate; thus in not-too-distance future, their supply of resources may become depleted.

Luckily for Black Africa, most of its resources remained obscured because they have been ignored by the super industrial world, for no reason whatsoever, except for the fact that the Whites have successfully convinced the world that everything about Blacks and Black Africa is inferior.

However, ignoring Black Africa and its resources might be the best thing that ever happened to the Black race because these resources will definitely explode into booming world market in the beginning of this twenty-first century.

In this chapter, we will examine the availability of these resources in order to accurately predict their potential economic impact and benefits to the Black man, as he struggles to attain superior economic power in the twenty-first century.

Let us begin this complex examination with the mineral resources because mineral resources are currently the backbone of certain Black African countries' economy.

Black African Mineral Resources:

Black Africa is probably one of the wealthiest in mineral resources in the world today. Since petroleum is one of its mineral resources of interest to the industrial world, especially the United States, we will first examine Black African petroleum.

The world community now realizes that oil is being produced and consumed at such a tremendous and ever-increasing rate that in not-too-distant future, the world oil supply could become depleted.

The uncertainty of adequate oil supply has necessitated an urgent need to explore advance scientific and engineering approaches necessary to develop alternative energy sources to promote energy conservation and achieve a healthier environment. However, this advanced scientific and engineering approach to develop alternative energy sources is at the preliminary stage, and no one can predict with certainty the impact of such effort on the world oil market.

Nigeria, with its oil refineries in many parts of the country, is probably the only documented country in Black Africa that produces and exports oil. How about other Black African countries, such as Cameroon, Togo, Ghana, Ivory Coast, Liberia, Sierra Leone, Senegal, Kenya, Uganda, and Ethiopia? According to the research and survey conducted by the Young African Engineer Association (YAEA), promising scientific data exists to predict the presence of high volume of drillable oil in many parts of these countries. The study also revealed promising scientific data to believe the presence of high volumes of other marketable minerals in many parts of these countries. These minerals include

- ✓ coal and gold
- ✓ diamond and uranium
- ✓ lead and zinc
- ✓ limestone and tin
- ✓ columbite and iron ore
- ✓ marble and stone
- ✓ zircon and petroleum

YAEA consists of American- and European-trained medical personnel and engineers from the following sub-Saharan African countries:

- ✓ Cameroon and Nigeria
- ✓ Togo and Ghana
- ✓ Ivory Coast and Liberia
- ✓ Sierra Leone and Senegal

✓ Kenya and Uganda
✓ Ethiopia

These untapped mineral resources is expected to attract numerous investment opportunities and definitely explode into one of the largest world market in history in the early stage of this twenty-first century, giving Blacks superior economic edge over other races for the first time in modern history.

	Liberia	Ghana	S. Leone	Nigeria	Senegal	Togo:
Coal				x		
Zinc	x	x	x	x	x	x
Gold	x	x	x	x	x	x
Diamond		x		x		
Uranium	x	x		x		
Petroleum				x		x
Limestone	x	x	x	x	x	x
Tin	x	x	x	x	x	x
Columbite	x	x	x	x	x	x
Iron Ore	x	x	x	x	x	x
Lead	x	x	x	x	x	x
Marble	x	x	x	x	x	x
Stone	x	x	x	x	x	x
Zircon	x	x	x	x	x	x

Table 5: Untapped Sub-Saharan African Minerals
Resources

	Kenya	Uganda	Ethiopia	Cameroon:
Coal	x	x	x	x
Zinc	x	x	x	x
Gold	x	x		
Diamond			x	x
Uranium	x	x		x
Copper			x	x
Petroleum	x	x		
Limestone	x	x	x	x
Tin	x	x	x	x
Columbite	x	x	x	x
Iron Ore	x	x	x	x
Lead	x	x	x	x
Marble	x	x	x	x
Stone	x	x	x	x
Zircon	x	x	x	x

Table 6: Untapped Mineral Resources of Other Parts of Black Africa, including Cameroon in West Africa

Black African Agricultural Resources

Can the Black African agricultural resources be another recovering weaponry against the economic challenges confronting the Black race? From every angle of analysis, Black Africa is a blessed land. Located along the equator with very favorable climate all year round, Black Africa has enormous advantages of growing and producing more healthy agricultural resources when compared to other parts of the world.

Due to rapid population growth in the industrial world, the consumption of agricultural resources has hit a peak and is on the way down, creating glut in the world food market. The anticipated food shortage associated with rapid population increase necessitated the desire to grow agricultural resources with artificial ingredients to meet the increasing demand.

Subsequently, the artificial ingredients create complex health problems to a large percentage of the population. Hence, there is an urgent need to produce and supply more healthy food products, which means natural food products with no artificial ingredients whatsoever. To meet the increasing demand with natural food products, especially in the United States, will pose some difficulties without additional outside assistance. This could certainly deplete the industrial world food supply.

Hence, Black African agricultural resources can play a pivotal role in creating sustainable and dependable world food markets with its healthy natural food products. The research and survey conducted by the Young African Engineer Association (YAEA) revealed that the following Black African agricultural resources exhibit superior nutritional characteristics.

These agricultural resources include

- ✓ fruits
- ✓ nuts
- ✓ banana

✓ plantain
✓ various types of vegetables
✓ rice
✓ yam
✓ cassava
✓ meat
✓ seafood
✓ palm
✓ Oil

We will employ tables 7 and 8 to demonstrate the production and distribution of these Black African agricultural resources.

	Liberia	Ghana	S. Leone	Nigeria	Senegal	Togo:
Fruits	x	x	x	x	x	x
Nuts	x	x	x	x	x	x
Banana	x	x	x	x	x	x
Plantain	x	x	x	x	x	x
Vegetable	x	x	x	x	x	x
Rice	x	x	x	x	x	x
Yam	x	x	x	x	x	x
Cassava	x	x	x	x	x	x
Meat	x	x	x	x	x	x
Sea Food	x	x	x	x	x	x
Palm Oil	x	x	x	x	x	x

Table 7: West African Superior and Nutritional
Agricultural Resources

	Kenya	Uganda	Ethiopia	Cameroon
Fruits	x	x	x	x
Nuts	x	x	x	x
Banana	x	x	x	x
Plantain	x	x	x	x
Vegetable	x	x	x	x
Rice	x	x	x	x
Yam	x	x	x	x
Cassava	x	x	x	x
Meat	x	x	x	x
Sea Food	x	x	x	x
Palm Oil	x	x	x	x

Table 8: Superior Black African Nutritional Agricultural Resources of Other Parts of Black Africa, including Cameroon in West Africa

Black African Natural Resources

Tropical natural products are possible sources of drugs against malaria, parasitic diseases, diarrheal disorders, infectious diseases—including AIDS and its opportunistic infections—cardiovascular diseases, respiratory diseases, hepatitis, mental disorders, and other serious illnesses prevalent to man.

As a result, many biochemical developments and pharmaceutical companies have been involved in screening and analyzing natural products for antitumor and AIDS-antiviral activities and have tested over 114,000 extracts from over thirty-five thousand plants. Interestingly, over 75 percent of these plants come from Black Africa. In spite of this high percentage, a large number of native Black African herbs (NBAH) have not been widely utilized in either biomedical research or the biotechnology market, creating additional economic frustration to the Black race.

According to the ancient Black African herbalists (ABAH), the universe is composed of opposing, but interdependent forces. Interestingly, this philosophy resembles the concept of homeostasis, the natural balance that occurs within living organisms, including the harmony between antagonists and agonists that regulate vital functions. Thus an important factor in the search for new medicine is developing compounds that work together with the body's own restorative and regenerative abilities.

To lead healthy lives, we must seek balance with nature, with society, and within ourselves. Through the medicinal and scientific utilization of native Black African herbs (NBAH), the Black race can make significant contribution to humanity toward achieving such balance.

Unfortunately, however, native Black African herbs (NBAH) remained obscured to the world communities due to the following:

✓ lack of documented proof by scientific communities
✓ lack of education by primitive African senior citizens
✓ lack of knowledge in pharmaceutical science
✓ lack of knowledge in radiation physics
✓ lack of knowledge in nuclear technology
✓ lack of adequate communication with outside world
✓ prevalent poverty
✓ lack of modern technology and resources

No scientific investigation has ever been encouraged or undertaken to explore NBAH as a cost-effective alternative medicine against AIDS, cancer, and other sicknesses of importance to the United States and Black Africa. In addition, NBAH could become a biotechnology cornerstone that can provide several important scientific and economic benefits to the United States and Black Africa including

✓ biotechnology-related investment opportunities
✓ profitable business relationship
✓ wide utilization of NBAH as effective alternative medicine

Natural products most promising for drug development are often found in ecosystems that are seriously threatened. These include tropical herbs, such as native African herbs (NAH).

The terrible irony is that as advances in biology expand our ability to use genetic diversity for drugs development, the raw materials are being lost to extinction.

The loss of cultural diversity threatens traditional knowledge developed over generations to identify plants and animals with medicinal value. The underlying causes of this biodiversity loss can be attributed to interwoven social, economic, and political elements. Poverty, unemployment, and lack of economic opportunities are also significant contributing factors.

Efforts to protect biological diversity in Africa will succeed, if implemented with the involvement of Black Africans and the understanding that Blacks' economic prosperity must originate from Black Africa, if it is expected to make any significant contributions and lasting economic impact. The research conducted by BTA confirms that native African herbs (NAH) contain a wealth of potentially useful compounds from which over 75 percent of biologically effective drugs have been isolated.

If these findings about NBAH are indeed true, the associated scientific merits and technical feasibilities can be very enormous with virtually unlimited potential payoffs that can endear its proponents to mankind forever.

The successful isolation, characterization, and biological testing of compounds from NBAH could generate scientific data needed to advance the pharmacologically value of NBAH and identify Blacks for the first time in modern medicine with science- or technology-related products.

Recognized Effectiveness of NBAH:

We will identify four outstanding Black African herbs with remarkable healing characteristics, by briefly discussing their background, history, and their treatment capabilities. These three herbs are currently being studied at BTA. (If you are interested to know more about these NBAH, please contact BTA at 1190 Saratoga Avenue #150, San Jose, California 95129 [408] 244-7852, btagrant@comcast.net)

Ethiopia Sarcophyte piriei:

This Native African Herb (NAH), Sarcophyte piriei Hutch, (Balunophoraceae; vernacular name, diinsi) is a parasitic plant that grows on the root of Acacia species. A decoction of its underground tube is in Ethiopia, Africa. It is a popular folk remedy against bruises, toothache, sore throat, and abdominal pain.

Nigerian (Igbo plant) *Phyllanthus niruri*:

This Native African Herb (NAH) Phyllanthus niruri L (Euphorbiaceae) is an Igbo plant. Igbo is a tribe in Eastern part of Nigeria. This NAH has been widely used against jaundice in Igbo traditional medicine and also for jaundice in Indian traditional medicine.

Nigerian (Igbo Plant) *Clitoria ternatea* Flowers:

This Native African Herb, Clitoria ternatea L, (Leguminosae), butterfly pea in English and omiro in Nigerian (Igbo language) has bluish purple flowers. It is used as food colorants in Nigeria (Igbo land) and Southeast Asia.

While these NBAH exists only in Nigeria and Ethiopia, numerous other NBAH exists all over Black Africa that is possible sources of drugs against

- ✓ malaria
- ✓ parasitic diseases (PD)
- ✓ diarrheal disorders (DD)
- ✓ infectious diseases (ID)
- ✓ AIDS and its opportunistic infections
- ✓ cardiovascular diseases (CD)
- ✓ respiratory diseases (RD)
- ✓ hepatitis mental disorders (MD)

We will employ tables 9 and 10 to demonstrate where the NBAH exhibit superior healing characteristics against the illness listed and where found in Black Africa.

	Liberia	Ghana	S. Leone	Nigeria	Senegal	Togo
Malaria	x	x	x	x	x	x
PD	x	x	x	x	x	x
DD	x	x	x	x	x	x
ID	x	x	x	x	x	x
AIDS	x	x	x	x	x	x
CD	x	x	x	x	x	x
RD	x	x	x	x	x	x
Hepatitis	x	x	x	x	x	x
MD	x	x	x	x	x	x

Table 9: Sicknesses Cured with Native Sub-Saharan
African Herbs

	Kenya	Uganda	Ethiopia	Cameroon:
Malaria	x	x	x	x
PD	x	x	x	x
DD	x	x	x	x
ID	x	x	x	x
AIDS	x	x	x	x
CD	x	x	x	x
RD	x	x	x	x
Hepatitis	x	x	x	x
MD	x	x	x	x

Table 10: Sicknesses cured by other native Black African
herbs, including Cameroon in West Africa

NBAH as Radiation Treatment

After the Second World War, the superior striking capability of nuclear weapons stimulated increased testing. The French government conducted numerous testing at African Sahara.

As with any nuclear blast, electrons emitted from the radioactive isotopes of the nuclear blast mixed with atmospheric dust were scattered throughout some parts of Nigeria. Nigerians in the effective area experienced the following radiation effects:

stochastic—This is function of dose without threshold, such as neoplastic disease (late effects).

nonstochastic—This is when the severity of effects depend on dose and for which a threshold may occur, such as early effects (fatigue, nausea, vomiting, etc). The treatment of black soap against this widespread of radiation sickness is provided in table 11. The number of deaths dropped significantly as more people became aware of the protective power of black soap around 1970. Black soap is a Black African soap. The soap is derived from Native African Herbs.

Table 11. *Black Soap* treatment against radiation poisoning as documented by Nigerian clinical registry

TEST DATE	ROENTGEN	EXPOSED	# OF DEATHS
1966	450-500	25,000	1683
1967	500-600	28,470	1042
1968	600-650	27,400	642

| 1969 | 600-675 | 20,910 | 320 |
| 1970 | 675-700 | 34,547 | 104 |

If indeed black soap was effective against beta and gamma radiation, which has worse damaging effects to the body cells than cancer or the AIDS virus, it is very possible, in addition to the black soap being possible countermeasures to AIDS and cancer, that it may contain the following:

✓ thiol radioprotector
✓ transdermal free radical scavenger
✓ surfactants and metal chelating

These unique characteristics of black soap, can qualify the soap as an effective treatment against radiation associated with the following:

UVB Radiation

Stratospheric ozone layer over the Antarctic has declined sharply over the last decade, and the depletion is expected to continue. This depletion has allowed enhanced penetration of UVB radiation to the earth's surface where UVB radiation has been historically low.

Even a small increase in ambient UVB radiation can cause serious skin cancer and other related skin problems. Black soap, with its pharmaceutical characteristics, can further be cost-effective treatment and protection against the skin cancer and other related skin diseases associated with this phenomena.

Space Radiation

Radiation exposure may not currently pose a significant problem on a short duration and low earth orbit of the space shuttle running about 5-10 milliards per day. This situation is likely to change as preparations for space station escalate.

Comprehensive radiobiological review of the physical interactions and transport of space radiation (protons, electrons, and galactic heavy ions) show that high-energy ions (HZE particles) posses unique radiation damage characteristics and may be highly carcinogenic for prolonged manned space mission. This is of a particular concern as increasing number of astronauts will be exposed to the radiation field existent in an orbit of about 450-500 km altitude and an inclination of about 28.5 degrees for extended period of time.

The radiation shielding or absorbing agents of the black soap has been of a variety of applications. Based upon its extraordinary protective measures against nuclear radiation, it can be very advantageous to adequately evaluate agents of black soap as cost-effective alternative medicine against biological effects of ionizing radiation.

Bone Demineralization

The biological implications of bone demineralization resulting from aging, prolonged weightlessness, and radiation effects have not been well investigated; and as a result, no effective countermeasures have been developed. For many centuries, physicians used prolonged bed rest and immersion in water as treatment for various diseases, especially in Black Africa where access to medical technology is lacking.

These treatments had positive remedial effects. However, adverse physiological responses become evident when the patients returned to their daily activities. Studies on these changes and their effects indicate

Bed Rest: Abrupt changes in body position cause acute changes in fluid compartment volumes, such as:

- ✓ fluid shifts and body composition
- ✓ renal function and diuretics
- ✓ calcium and phosphorus metabolism
- ✓ orthostatic tolerance and bone demineralization

Water Immersion: Abrupt changes in body position cause acute changes in fluid compartment volumes, such as:

- ✓ fluid shifts and body composition
- ✓ cardiovascular-respiratory responses
- ✓ natriuretic and diuretic factors
- ✓ renal function and bone demineralization

A significant percentage of Africans were using butter derived from native Black African herbs (NBAH-Butter) as treatment against fatigue and muscle weakness (prior to the introduction of prolonged bed rest) and water immersion as treatment against various diseases.

Using radiographic and clinical evaluations, physicians determined that among the Africans using NBAH-Butter prior to the bed rest and water immersion, no bone-demineralization-associated symptoms occurred, which was in marked contrast to those not using NBAH-Butter.

Another African bone-demineralization-related experience was during the French nuclear testing of African Sahara. The radioactive particles from these testing emit beta and gamma particles, which penetrate into the body tissues to damage the calcium minerals of the bones.

No topical treatment can reduce gamma radiation to the bones, however, the chemical characteristics of NBAH-Butter may have proved otherwise. The nuclear testing also ingested certain radionuclides into the food chains in Ghana. Ghanaians used NBAH-Butter as an effective treatment

CHAPTER 5

Blacks Compared against Other Ethnics

All technological or scientific advancements today were based upon the early research and related developments. For an example, the advancements in aviation technology today are all based upon the early research and development related efforts by the Wright brothers. Similarly, advancements in automobile technology are based upon the early research and development efforts by Ford. In parallel with this viewpoint, all the advancements in medical technology will also be based upon the early research and development-related efforts, which excluded Black race. Given this scenario, it is safe to conclude that the early exclusion of Blacks from medical or medicinal-related efforts had significant impact on effective treatment or cure among the Black population. This is because every medical device is a derivative of the early medical research and developments, which excluded Blacks.

All chronic conditions along with their associated serious illnesses have all been equated to the Black race. How accurate are these assessments? In order to perform a convincing analysis, we will compare sub-Saharan African nations against that of the industrialized nations. In so doing, we will be able to evaluate or reevaluate the natural resources of sub-Saharan African nations, particularly their agricultural resources against that of the industrialize nations.

The location of sub-Saharan Africa, along the equator, is a big advantage given the favorable climate with sufficient rainfall all year round. It is therefore safe to say that sub-Saharan African nations has the advantage of growing and producing more healthy-food-related products compared to that of the industrialized world.

The consumption of food-related products in the industrialized world, including the United States, has hit a peak and is on the way down due to rapid population increase. This is creating a glut in the world food market. The anticipated food shortage associated with this phenomenon necessitated the desire to grow unnatural food-related products to meet such growing demand. Subsequently, the counterfeit ingredients used for such food processing create chronic conditions and various other serious illnesses that bring higher percentage of the growing population into medical care. A large number of the population requires hospital treatment and many of them die from their illness.

Among those hospitalized, a large number receive intensive care, and the cost of such treatment runs to several hundred million dollars a year thereby compromising the health-care budget.

There is therefore an urgent need to produce and supply more healthy food products, which means natural food products with no artificial ingredients whatsoever. To meet such demand, especially in the United States, is expected to pose serious financial burden to the low-income working families. It could further deplete the industrialized world food supply.

Given these scenarios, one can now factor the sub-Saharan African nations agricultural resources into the food supply equation to determine the pivotal role it will play in creating sustainable and dependable world food supplies with healthy and natural ingredients.

Without adequate agricultural tools and machinery investment, it will be very difficult for the sub-Saharan African farmers to capitalize on this specific golden opportunity, which did not exist for them several decades ago.

Such opportunity is much needed at this time when the sub-Saharan African nations are at both social and economic crossroads.

The purpose of this brief discussion on natural and unnatural food products is to demonstrate how the sub-Saharan African natural food products could be responsible for a variety of health advantages that Blacks enjoy today. This is particularly very true with Black Africans who enjoy healthy advantages more than other ethnicities. Because of the quality of Black Africans' natural food products, Black Africans enjoy the following health related advantages:

- ✓ younger looking when compared to other ethnicities
- ✓ none or limited hospital visits in a lifetime
- ✓ aging gracefully
- ✓ longer life span (ninety to a hundred years old)

As a matter of fact, my grandmother never visited any hospital during her lifetime, yet she passed the age of ninety. Most native Black Africans, especially those in the rural areas, enjoy a life span of over one hundred years. This superior health advantage can be attributed to Black African diet. Black Africans' superior diet enables them to have plenty of energy and the ability to think on their feet.

These factors notwithstanding many Black Africans are still experiencing the following social and economic problems:

- ✓ starvation and malnourishment
- ✓ poverty
- ✓ underemployment
- ✓ poor access to the health care system
- ✓ weak economy

Part of these problems can be attributed to limited skills and tools required for mechanized farming. It is only through mechanized farming that Black Africans can be able to achieve mass production of their agricultural products. Without mechanized farming, production will be maintained at a local level, which could compromise production and affect the economic output.

Can Blacks enjoy quality health care in the twenty-first century after being medicinally and medically isolated and completely suppressed or oppressed for over one hundred years?

Before colonization, Blacks were physically fit due to the associated health advantages discussed earlier. The colonization of sub-Saharan African nations marked a miserable turning point to Blacks' overall conditions and economic conditions.

Given the sub-Saharan Africans' bows-and-arrows-defense capabilities against the colonizers, they were forced to disregard the use of both their natural resources including their traditional remedies and embrace the medicine brought along by the colonizers. In so doing, they killed the sub-Saharan African civilization by excluding Blacks from every decision-making process including their exclusion from any medical—or medicinal-related research and development efforts.

Also every pharmaceutically derived medicines or drugs were, in the same manner, derivatives of the early pharmaceutical research or drugs developments. These assessments make it very difficult for any physician or doctor, regardless of qualifications or experiences, to perform a satisfactory diagnostics and accordingly prescribe effective medication or treatment with any medical certainty on a Black person without trial-and-error techniques. I have quite often wondered why has no Black person come forward to challenge most of these baseless and unverifiable medical findings against Blacks; after all, Blacks do have qualified doctors and physicians.

I have wondered at a great length about this specific scenario. After a careful examination of facts and logical analysis, I came to realize that regardless of the magnitude of education and skills acquired by these Black scientists and doctors, they cannot be able to challenge any of these medicinal or medical deficiencies against Blacks. This is mainly due to the fact that their acquired skills and education were still and solely based upon the early research and development that completely excluded all Black population. It will therefore be very difficult if not impossible for any Black scientist or doctor to convincingly argued about this situation, considering that their education were based upon the medical and medicinal conclusion that did not include any Black population.

As a result, Blacks, regardless of economic status, will continue to accept whatever their medical doctor wants them to believe.

With the information available so far, one can make intelligent assessment of the accuracy or inaccuracy of the findings that are being addressed in this book.

The notion that Blacks have high blood pressure, particularly Black men, is still open for more scientific investigations. My argument is based upon the fact that the medical data from which such medical information were derived completely excluded any Black population from the preliminary research and study efforts from which the equipment was developed.

Without the inclusion of Black population in the preliminary research and development, no one, regardless of the level of medical prestige, can document with any medical or scientific certainty any of the medical or medicinal disorders attributed to Blacks all over the world.

Are these medical disorders against Blacks based upon a scientific finding, or is it a strategic plot to further intimidate and humiliate the Black race? Either way, there is no scientific data to justify such existing claims.

I once confronted a medical professional at a Sunnyvale clinic with this specific question. I asked this highly and well-respected medical specialist about her thoughts on Blacks and high blood pressure as it is currently claimed and documented in the medical literature.

This senior-level medical specialist told me in a very simple term that to her best knowledge, high blood pressure is related to unknown mechanism for conservation of water in hot climate. She continued because large part of Black Africa is located along the equator, it was believed that Blacks conserve high volume of water in hot climate and therefore must exhibit symptoms of high blood pressure.

She concluded by saying that based upon these facts, high blood pressure was attributed to skin color. She said probably that is the more reason why the medical community reached its findings and conclusions that Blacks, particularly Black men, have high blood pressure.

Is this the proof? You can answer the question better than I can.

If you really go through key medical findings and see how such findings were made, you probably will be shocked.

The fact of the matter is that indeed no accurate medical diagnoses currently exist to justify the findings that Blacks, particularly Black men, have high blood pressure and are more likely to suffer heart attack.

We will zero in on other disturbing medical findings against Blacks in this book to demonstrate that the Black population has been misled for a long time by the medical community.

By the way, is high blood pressure inborn or acquired through the type of environment one is constantly exposed to?

If high blood pressure is not inborn, then all the Black African slaves to this country did not exhibit any symptoms of high blood pressure upon arrival to this country. Were the Black African slaves diagnosed with high blood pressure upon arrival to the United States? I think not. Were Black Africans living in Black Africa diagnosed with high blood pressure at the time of the slave trade? Again, I think not.

I will repeat this very paragraph from previous chapter to enable me to bring my point across more effectively. Located along the equator with very favorable climate all year round, Black Africa has enormous advantages of growing and producing more healthy agricultural resources compared to other parts of the world. As a result, Black Africans eat natural healthy food products unlike the United States and other industrialized countries.

Could it be the artificial ingredients in the food products creating or causing the high blood pressure? Even if it is, still the notion that Blacks are mostly associated with high blood pressure is still questionable. The primary problem with Blacks' medical treatment is that every finding can be very questionable because the medical research from which all the medical decisions were based upon today were conducted without any statistical data from the Black population. The blood pressure reading system is not an expectation. How then can such instrument be used to accurately read the blood pressure of Blacks without any errors or mistakes?

One could therefore argue, if it works for other ethnicities, why not for Blacks. The answer is simple. Because the biological composition of Blacks and that of other ethnicities differ considerably, otherwise we all could just been one color.

Can misdiagnoses create serious medical problems? You bet, it can because once a doctor tells you that you are sick, you start immediately to feel sick whether or not you are sick. That is Physiology 101, and such experience can certainly intimidate someone.

This book has been written to argue that Blacks are not well represented in the medical community.

CHAPTER 6

Opposing Motion against the Black Race

In this chapter, we will examine the level of knowledge, skills, expertise, and apparatus available to Blacks to successfully develop appropriate technologies from their current available resources and commercialize the technological derivatives in the most profitable manner. Based upon the results of the examination, will an educational program be alternative approach to enhance critical-thinking, problem-solving, and decision-making skills of Blacks in terms of participating in technological-related undertakings?

Black African Mineral Resources

The major challenge that will face the Black African mineral industry will be the ability to respond to a highly volatile and fragmented market demand by rapidly launching new technological derivatives from these resources, in a manner that will be very profitable, and gain them science and technology advantages against their limited knowledge in developing attractive scientific products.

The type of expertise and knowledge needed for this complex undertaking include

metallurgical and materials engineering (MM Eng)
mining engineering (M Eng)
petroleum engineering (P Eng)
chemical engineering (C Eng)
geologists and geodesists (GG)

Table 12 has been compiled from the United States Housing and Household Economic Statistics Division, Department of Commerce, to compare the qualifications of Blacks against the scientific know-how required to launch this new technological derivatives from these resources.

	General	White	Black	White%	Black%
Population	267,000,000	219,600,000	33,600,000	82%	13%
MM Eng	19,000	17,000	665	90%	3%
M Eng	6,500	6,000	68	94%	1%
P Eng	24,000	23,000	539	93%	2%
C Eng	64,300	57,000	2,200	89%	4%
GG	53,000	51,000	604	96%	1%

Table 12: The scientific and technological capabilities of the White man compared against that of the Black man in the United States in developing associated technologies from the Black African mineral resources.

The figures on table 12 show that of 19,000 metallurgical and materials engineers (MM Eng) in the United States, only 665 (3 percent) are Blacks compared to 17,000 (90 percent) of Whites. Of 6,500 mining engineers (M Eng), only 68 (1 percent) are Blacks, compared to 6,000 (94 percent) of Whites. Of 24,000 petroleum engineers, only 539 (2 percent) are Blacks, compared to 23,000 (93 percent) of Whites.

Of 64,000 chemical e (C Eng), only 2,200 (4 percent) are Blacks compared to 57,000 (89 percent) of Whites. And of 53,000 geologists and geodesists, only 604 (1 percent) are Blacks compared to 51,000 (96 percent) of Whites.

Agricultural Resources

The ability of the Black Africa to respond to highly volatile and fragmented market demand, in the highly competitive global marketplace, in a profitable manner will require improved knowledge in science- and technology-related subjects.

This may not be that challenging because the Black man has been involved with agricultural cultivation since the beginning of time. However, the cultivation has been on a smaller scale. Expanding agricultural production to a large scale will require additional expertise and know-how. The type of experts and specialists needed for this complex undertaken will include

✓ agricultural and food scientists (AF Sci)
✓ biological and life Scientists(BL Sci)
✓ forestry and conservation scientists (FC Sci)
✓ agricultural engineers (A Eng)

Table 13, (compiled from the United States Housing and Household Economic Statistics Division, Department of Commerce) compares the qualifications of the Black man against the scientific know-how required to successfully make Black African agricultural resources one of the main export products.

	General	White	Black	White%	Black%:
Population	267,000,000	219,600,000	33,600,000	82%	13%
AF Sci	35,000	32,000	1,500	91%	4%
BL Sci	62,000	54,400	2,4003	88%	4%
FC Sci	34,800	32,500	1,100	93%	3%
A Eng	2,400	2,000	45	93%	2%

Table 13: The scientific and technological capabilities of the White man compared against that of the Black man in the United States in developing Black African agricultural resources into a large export market

Natural Products

The major challenge that will face Black Africa pharmaceutical and biotechnology industry will be the ability to convince the world health care industry that Black African natural products, such as native Black African herbs (NBAH) exhibit remarkable healing characteristics that can justify the necessary investment in the research and development of pharmaceuticals from these herbs into the world health care market.

Certainly, it will be difficult for Blacks to successfully carry this burden, considering their limited expertise in this domain. Does this then mean that the desire of Blacks to gain superior economic status in the twenty-first century is defeated? Definitely not!

This simply means that Blacks will face challenging and complex problems on their path to economic revolution.

How will Blacks tackle these challenging and complex problems? Two answers might apply. One answer, which is not to the best business interest of Blacks, will be to invite international investors to invest heavily on the research and development necessary to successfully develop and market biomedical and biotechnological products, resulting from these resources, in the most competitive manner.

This is probably bad for Blacks because they will still depend on foreign investors and competitors who will take over 50 percent of the associated profit plus other associated charges.

The other answer might be through comprehensive educational programs. It is only through education that Blacks will be able to acquire the necessary skills and expertise required to develop associated technologies from their resources.

How to develop such educational programs to meet the scientific and technological needs will pose additional complications to the program development effort. Black Technologies Advancement has developed competitive advantage strategies to help Blacks to challenge this important problem in dealing with competitive skills in high-technology industry.

Today, Black African minerals are fragmented, each country having a unique set of researching, internal data formats, availability of resources record systems, and communication and computer networks. This lack of an adequately developed infrastructure is a cost barrier that will pose additional problems to the development effort.

There are also high technical risks in the systems engineering to ensure reliability, availability, maintainability, data integrity of these resources and the high level of confidence needed to derive the associated technologies from these resources in the cost effective manner.

This program at Black Technologies Advancement is designed to provide evolutionary changes and advances in information technology as applied to Black African resources that will change and improve the economic status of the Blacks.

CHAPTER 7

Blacks Association with Serious Diseases

The White population, do have lower number in all disease categories. Of course, they should because all the early medical research and developments were based upon their race and their population.

Any serious or life threatening illness or diseases of major concern to society are usually attributed to Blacks. A good example is documented by African American Health Facts (the Office of Minority Health Resource Center), Washington DC in 1997.

This document emphasized, that according to the National Center for Health Statistics, "Advance Report of Final Mortality Statistics: Monthly Vital Statistics Report," Vol. 45 No. 3(S), September 30, 1996, the top-ten leading causes of death among African American men and women combined in 1994 were as follows:

Blacks' association with other serious illnesses

1. heart disease—27.2 percent
2. cancer 21.2—percent
3. stroke 6.4—percent
4. HIV infection—5.7 percent
5. unintentional injuries—4.5 percent
6. homicide—4.3 percent
7. diabetes—3.5 percent
8. pneumonia/influenza—2.6 percent
9. chronic obstructive pulmonary diseases—2.3 percent
10. perinatal conditions—2 percent

These particular facts were further well documented by the Empowerment Initiative (TEI). TEI is a bimonthly newsletter geared toward empowering African Americans.

In addition to the ten diseases listed, Black women are now very much associated with breast cancer.

According to the recent medical literature, Black women are more likely to suffer from breast cancer than other ethnic groups put together.

Furthermore, cancer has been documented as the second leading cause of death for Blacks with a higher age-adjusted death rate than any other racial group.

The statistical breakdown reads as follows: Between 1973 and 1991, cancer mortality for white males increased by only 5.6 percent, while it increased by 24 percent for Black males.

For females, the rate increased by 13.3 percent for Blacks and only 8.1 percent for Whites. How were these statistical data generated?

It is undocumented medical fact that Blacks usually do not suffer from melanoma.

Otherwise, Black Africans would be dying every second of the day from melanoma resulting from the intense high temperature from the sun due to the location of Black Africa along the equator.

Is there any scientific or medical proof to justify this notion that Blacks are more likely to suffer from cancer than other races? If there is one, on what medical logic was the study based upon? What equipment, instrument, or medical tools were used for the study? Were testing involved? If so, what tools, equipment, or apparatus were used to conduct such tests?

How were these equipment, tools, instruments, apparatus, medical systems, or devices developed? Was any Black population included in any of the preliminary studies or research from which these medical systems or devices were developed?

These should be the questions Blacks should ask themselves whenever they are diagnosed with any of these Blacks-associated illnesses. It is very disappointing that Blacks agree and accept these medical findings without any challenge of the data.

While studies indicate that many African American population are at high risk for HIV infection, not because of their race or ethnicity, but because of the risk behaviors they may engage in. We all know at firsthand that teenagers and youths of poor ethnic minority parents are subjected to serious economic problems that may force most of them into unsafe sex behaviors due to constant use of drugs and alcohol.

They are in constant use of alcohol and drugs, because there is nothing else to occupy their time. Once, such tragic scenarios exit the drug-abuse problems as well as unsafe sex behavior become very rampant.

Drugs and alcohol are the triggering factors for unsafe sex behaviors. Unsafe sex behaviors are the primary cause of AIDS/HIV virus.

As with any population, it's not who you are but what you do that puts you at risk for HIV. It has been documented that African Americans are disproportionately affected by HIV as compared to other races because African Americans account for 33 percent of the total AIDS cases in the United States while comprising only 13 percent of the United States population. This documentation continued to indicate that in 1994, more than half (57 percent) of AIDS cases among women were among African Americans. Likewise, African Americans accounted for over half of all AIDS cases among injection drug users (IDUs). In 1994, 62 percent of all children with AIDS were African Americans.

What the study did not say is that a wide variety of populations in the United States were included under this heading. Lower class, Christian, Muslim, inner city, suburban, descendants of slaves, and recent Caribbean immigrants all came under the African American heading. Why did the study fail to record these social, cultural, economic, geographic, religious, and political differences so that accurate percentage for Black population will be documented accurately?

It is established fact that Blacks are affected by AIDS/HIV virus in a high number, but it is unfair to group them with other groups to make them think otherwise.

Scientists in Black African countries have made impressive headway in developing effective countermeasures against AIDS/HIV virus. However, they have continued to experience major difficulties in developing these countermeasures due to lack of funding.

United States Vice President Al Gore was recently challenged by Black African AIDS/HIV virus advocates during his presidential campaign kickoff in his hometown.

This was because he had failed to show any support to generate any type of funding for Black African AIDS/HIV virus research. In 1993, local, state, and territorial health departments reported to CDC 58,538 cases of AIDS among racial/ethnic minorities.

A total of 38,544 (66 percent) cases were reported among blacks, 18,888 (32 percent) among Hispanics, 767 (1 percent) among Asian/Pacific Islanders, and 339 (1 percent) among American Indians/Alaskan natives. White population were missing from this data.

We may employ this specific analogy to reevaluate the preliminary funding on high blood pressure, cancer, and other serious illnesses attributed to Blacks.

CHAPTER 8

Black Patients and Doctors

Total isolation of the Black population from the early medicinal or medical research has continued to devastate the Black race as a whole from satisfactory medical treatment or cure. To date, doctors, particularly White doctors, are still having difficult times providing quality or comprehensive health care services to Blacks.

Patient-doctor relationship and interaction of Blacks should be of concern to the medical community. In chapter 4, we established the fact that all serious and life-threatening illnesses are always attributed to Blacks. Yet no particular attempt is made to improve or enhance Black patient—doctor relationship. We all know that quality health care is the prerequisite for good health. Because when you are in good health, you can accomplish a lot with perfect peace of mind. Good health and peace of mind are the most important facts of life.

Before we proceed further with this chapter, we need to define a patient as it will be used in this book. In this book, particularly in this chapter, we will define a patient in the following manners:

✓ One is considered a patient in this book when he or she is admitted into a hospital for overnight or extended period of time.

✓ One is considered a patient in this book when he or she visits either his or her doctor or a doctor for a less-serious illness that can allow out-of-hospital treatment.

✓ Lastly, one is considered a patient in this book when he or she visits his or her doctor for regular medical checkup.

With these facts at hand, we will proceed to examine the relationship between a Black patient and a doctor. We will compare the result of our examination against that of a White patient—doctor relationship.

Based upon the result of the findings, we will draw our conclusion in favor of the most unfavorable relationship.

We will therefore proceed with our examination in the following manner:

Black patient—doctor relationship, in most cases, have been very unfavorable and needs to improve if Blacks can ever enjoy any improved health care benefits. Black patient—doctor relationship has been unfavorable due to the following various reasons, which could probably be explained in the following manners:

✓ In most cases, a Black patient—doctor relationship can be affected probably due to the doctor's general overview on Blacks. Regardless of how wealthy and successful a Black person may be, he or she is still a Black person. No one can change that simple fact. As a result, society, including doctors, look at Blacks in a certain way.

✓ In other cases, a Black patient—doctor relationship may be compromised probably due to the inability of Blacks to command any influence in the medicinal or medical undertakings. Such lack of influence can cause a doctor to look down on Blacks. This can result to a poor medical care on the Black patient.

✓ In some cases, a Black patient—doctor relationship may be jeopardized probably due to the fact that the doctor may not be well equipped with the necessary and important medicinal and medical data on a Black person.

Such information is needed to perform accurate medical assessment of a patient, in this case a Black patient, and correctly prescribe the most effective medication that will minimize or eliminate altogether trial-and-error techniques.

This is particularly true, especially when the sickness is unique and life threatening. The most compelling argument from these three viewpoints, which could be used to support unfavorable Black patient-doctor relationship, can be attributed to the last argument. This specific argument takes us back to the basic point of this book.

When a Black person suffers from unique type of illnesses, the chances that doctors will employ trial-and-error techniques to achieve effective treatment increases significantly. In such situations, the Black patient will accumulate additional illnesses or sicknesses due to the side effects experienced from such trials and errors with other medications. In other words, while the doctor is trying to achieve a cure, he or she might actually be worsening the Black patient's health.

At this point, we will examine a White patient-doctor relationship. We will compare the result of the findings as indicated at the beginning of the chapter.

The White patient-doctor relationship is the most favorable relationship. This is in part because the doctor, in most cases a White doctor, understands the biological makeup of his fellow White population. Furthermore, the doctor knows exactly how to perform an accurate assessment of his or her fellow White patient. In the cases where the doctor is of other ethnicity, the White patient will still enjoy favorable patient-doctor relationship. This is because while the doctor may be of other ethnicity, his or her medical or medicinal training and education were based upon the White population.

This is in part because all the early medicinal and medical research was all based upon the White population.

One can say, therefore, that White population enjoys every medical and medicinal advantages, not only in the United States, but also in other parts of the world.

CHAPTER 9

Blacks and the News Media

Why are Blacks not helping one another? This is a good question. Compare this question to another question, Why are Blacks not united? Again, another interesting question! Again, compare this very question to another question, Why are Blacks afraid to trust one another? And finally, Why is there so little teamwork and unity among the Blacks today?

The correct answers to these questions rest with the worldwide influence of the news media. The news media has succeeded to confuse Blacks to the extent that Blacks do not trust one another.

While studies have shown otherwise, the news media has successfully created a general impression among Blacks that they are inferior to other races. This inferiority complex has been with Blacks since the news media first launched its negative publicity against the Black race. Why the news media are against Blacks is not known. However, it appears that slavery and colonization might have been the influencing factor.

It has been said that birds of the same feather flock together. If this ideology can be applied to any race, it definitely will be most appropriate to the Black race. This is because of the problems created by the news media. The image of Blacks as created by the news media is very discouraging, especially to the Black youths. As a result, too many upcoming Black children want to be athletes or entertainers because that is all they see whenever they watch the television.

Frankly speaking, Black entertainers or athletes who are not knowledgeable in science, technology, mathematics, or computer science are not doing anything to address this particular problem among Blacks. As a result, many of them refuse outright to assist or support few Blacks associated or involved with science and technology.

We can now revisit one of the questions at the beginning of this chapter, Why are Blacks afraid to trust one another? This could be due to the limited knowledge of Black entertainers or athletes on science or technology. This might have been compounded by the negative image created by the news media against the Black race.

Consequently, Blacks with economic potential to create better economic environment for Blacks across the board refused to do so because they believe that few Black engineers and scientists will take advantage of their investments.

It is the belief of too many entertainers and athletes that the Black engineers or scientists will be dishonest with any monetary assistance given. This belief is based upon the limited engineering or scientific knowledge of the entertainers and the athletes. Subsequently, well-qualified Black engineers or scientists, especially those owning hi-tech small businesses, find it very difficult to generate any money from Black entertainers or athletes.

As a result, Black hi-tech companies with demonstrated successful track records continue to struggle to generate the necessary capital to expand businesses and create more employment opportunities. One of the ways to identify Blacks with science- or technology-related products in the twenty-first century will be through owning wealthy and successful hi-tech businesses and employing the resulting economic benefit to the development and marketing of Black African resources. It will be very difficult to identify Blacks with science- or technology-related products in the twenty-first century without teamwork between Black Africans and Black African Americans. Furthermore, it will be difficult to create a solid Black technology base without the successful solicitation of the support from the Black entertainers and athletes. It is sad that Blacks face numerous challenging science- or technology-related difficulties as we approach the twenty-first century.

The most frustrating scenario in this equation is that Black scientists or engineers with innovative ideas in science or technology are often unable to demonstrate the feasibilities of their ideas, which certainly hinders these innovations from ever making it into the market place. This is in part due to the inability of these Blacks to generate the necessary funding needed, not only to demonstrate the concept of their ideas, but also to actually develop the idea into a commercial product.

The challenges facing Blacks to be identified with a science- or technology-related product, as we approach the twenty-first century, is the ability to encourage upcoming Black children to break away from the orientation of Blacks' tradition of becoming entertainers or athletes and become engineers or scientists and team up with the growing number of Black engineers and scientists exploring how to best develop the Black African resources into the booming world market.

Creating financial opportunities for the current Black engineers or scientists will enable them to successfully demonstrate the concept of their ideas and successfully bring such ideas into a commercial product. How to create such opportunities for this group of Black engineers or scientists will be very difficult. The difficulties are associated with the complications and problems experienced by Black engineers or scientists in generating the necessary funding needed to expand their hi-tech small businesses.

Similar difficulties are experienced by the Black engineer or scientist to demonstrate the concept of an idea to the extent needed to bring such an idea into a commercial product. This, in conjunction with the negative publicity of the news media, further complicates the problem. A well-designed, mapped-out strategy or plan can help Black engineers or scientists to be launched into the twenty-first century.

Any effective or constructive plan to achieve this specific objective must involve, in some way, a grassroots education and a well-organized networking effort. Without any such strategy, complications will develop.

Chapter 10

Will Prayer Be Weaponry for Blacks Economic Power

According to the poem of the struggling boy from Black Africa, if any race will ever be commended for worshiping or praying, the Black race certainly needed to be commended. This successful track worshiping record notwithstanding, Blacks still faces serious and perhaps the worse economic problems in the world today.

Has anyone in the Black community ever wondered why this is the case for anyone from another race for that matter? If Blacks, in spite of their impressive worshiping records, still face the worse economic problems in the world today, they must not be worshiping according to the rules of nature or the rules set forth by God in this specific issue.

Several decades ago, the Black African Americans were experiencing the worse social and civil rights problems in the world, and Blacks in Black Africa were seriously suppressed by the Whites who colonized them. Blacks were worried and concerned about these sad situations and wanted to end the suppression outright.

The African Americans worshiped and prayed for an immediate end to this sad experience, so did the Black Africans. Because Blacks at that time prayed and worshiped with defined and specific goals in mind, God answered their prayers by giving them prominent leaders that successfully led them to end these tragedies. The leader for African American in this crusade was Dr. Martin Luther King Jr. God blessed Dr. King with all the necessary wisdom he needed to successfully lead this crusade. Despite the fact that he lost his life, his mission was accomplished. He gave Blacks the social equality they seriously deserved. The leader of Black Africa in his crusade was Dr. Nnamdi Azikiwe. His experience with the social injustice in the United States was very influential in his assuming the leadership role in Africa.

Dr. Azikiwe assumed the role to lead Black Africans out of the suppression imposed by the Whites who colonized Black Africa. Upon returning to Nigeria in 1958/59 after completing his education in the United States, Dr. Azikiwe decided to expel all the Whites who refused to depart from Black Africa after the colonization. His mission was a successful one because in 1960, he was able to secure independence for the biggest and richest country in Black Africa—Nigeria—from her British master; and Dr. Azikiwe became the first Nigerian president. In the subsequent years, other Black African countries followed suit and became independent as well.

While Dr. Azikiwe was the person that actually motivated and engineered Dr. King to lead the crusade in the United States, the name of Dr. King has remained immortal in history while that of Dr. Azikiwe remained obscured, especially in the United States and probably other parts of the world due to the type of coverage presented by the news media.

However, this chapter will not focus on these obscurities. Rather, it will focus on investigating why the Black race, up this moment, faces the worse economic problems in the world in spite of their aggressive and vigorous worshiping and praying power. Looking back, why was the prayer of Blacks to end the social injustice and suppression against them was answered? But since then the prayers to date to end Blacks economic struggle remained unanswered. Blacks succeeded as a team this one time because they were able to perform in the following manners:

- ✓ They were able to specifically identify their problem.
- ✓ They were able to specifically identify their plans to solve the problem.
- ✓ They were united as a team for a common cause.
- ✓ They had a burning desire for a specific goal.
- ✓ They performed as a team with designated leaders.
- ✓ They unite behind the leaders with common purpose.
- ✓ Finally they prayed for that defined purpose and succeeded.

Based upon these reasons, Blacks were able to end the social injustice and suppression with astonishing results in the quickest possible time. If the same techniques and methodologies applied in dealing with social injustice and suppression can be applied to the problem of poor economic status, the economic problems of Blacks would come to an end as well in the quickest possible time.

Unfortunately, however, no plans exist currently to challenge the problem of poverty among Blacks. The approach at the moment is to complain and complain. Realistically, no amount of complaints can resolve this problem. Rather it will make the problem worse.

No amount of complaining or frustration can resolve this economic problem without first trying to correctly identify the cause of the problem. If Blacks will ever be identified with science- or technology-related products, they must first be convinced that their economic problem originates from their inability to be identified with science- or technology-related endeavors.

Since this will be the first time such an argument is made, it definitely will be difficult to win a majority of supporters unless organized or directed by a designated leader knowledgeable on this very subject. This designated leader will not only be knowledgeable as to what science or technology will best suit the Black race but he will also be dedicated with a burning desire and unshakeable commitment to resolve the challenging economic problem of the Black race.

This unshakeable commitment will be measured by the extent this leader is willing to go in order to succeed. A major grassroots crusade will be undertaken by this designated leader to educate Black communities on the specific cause of action to resolve this devastating economic problem. The purpose of such a crusade will be to unite the Black race as a team for a common cause.

This team effort will clearly convince and motivate wealthy Black individuals to provide the necessary financial support to successfully launch this crusade. Until the Black race is united as a team and work for this specific cause and believe that the problem identification is correct, and the appropriate cause of action has been identified, it will be very difficult to succeed in this effort.

And again, until Blacks realize that without their involvement in technology or science, their economic security will remain in jeopardy; they will continue to feel inferior to other races for an extended period of time.

With these facts in place, the Black race can now pray or worship with a defined purpose to solve their challenging or complex economic problem. It is a known fact that prayers with no specified purpose usually go unanswered, but prayers with defined purpose are always answered. From any angle of analysis, you must know what you want before you ask for it. Without saying precisely what you want, there is nothing to give.

Part of the explanation of why all the prayers of the Black race have not been answered in relation to solving their economic problem is that no one in their community has ever tried to find out the reason why Blacks face the worse economic problems in the world today.

The reason why Blacks face the worse economic problems in the world today is because they are not identified with any science- or technology-related products. Science and technology equals economic security. This methodology of hope is the prerequisite for economic security without which any race will experience major economic problems for extended periods of time.

Therefore the identification of Blacks with science- or technology-related products are the prerequisite for their economic prosperity. There are several science- or technology-related products that are appropriate for Blacks involvements, but there are specific science- or technology-related products that will be most appropriate to give Blacks quick head start to achieve the following:

✓ superior economic power
✓ domestic and international prestige
✓ improved academic status in science and engineering

Appropriate science- or technology-related products must come from Black Africa. The Black African resources discussed in chapter 2 have all the necessary requirements needed to give the Black race the technology they will need to identify them with science- and technology-related products.

CHAPTER 11

Competitive Advantage Strategies for the Black Race

Based upon the information contained in the past chapters, one can see at a glance that Blacks lack the expertise and qualifications at this time to successfully develop associated technologies from their three key resources covered in chapter 2.

To assist Blacks with this problem, an educational program will be developed. The purpose of the educational program will be to assist them with the skills they will need to confront the challenging scientific and engineering tasks ahead.

- ✓ Biomedical research and manufacturing
- ✓ Engineering design and analysis
- ✓ Management and computing strategies
- ✓ Bioengineering

By stimulating commercialization of the technologies derived from Black African resources, Blacks will foster the integration of their defense and commercial industrial base with the world market while providing for the global competitive market the cutting edge in the commercial and defense capabilities.

The course outline for action plan implementation include

Biomedical research and manufacturing modules

Module 1—A Business Perspective on Manufacturing

- ✓ Review of Workshop Material
- ✓ Review of Business Basics
- ✓ Thinking Like Executives
- ✓ Description of Business Simulator
- ✓ Relationship to Workshop Objectives
- ✓ Homework Assignment

Module 2—Agile Manufacturing Strategic Importance of Manufacturing

- ✓ Manufacturing and Strategic Planning
- ✓ Competing in a Challenging Environment
- ✓ Integration of Manufacturing into the Enterprise
- ✓ Changing Paradigms
- ✓ Choices in Manufacturing Approaches/Structures/ Models
- ✓ Characteristics of Successful Manufacturing Organizations
- ✓ Integrated Product Teams, Concurrent Engineering/DFM
- ✓ Reengineering Products and Processes

✓ Information Technology Support to Manufacturing
✓ Cycle Time, Productivity, and Profitability
✓ Keys to Competitive Success

Module 3—Total Productive Maintenance

✓ Introduction to TPM
✓ The Basic Concepts
✓ Relationships to TQM, JIT, ABM, and TEI
✓ Roadmap for TPM Implementation: Awareness
✓ Organization, Planning, Implementation, Assessment
✓ Equipment Losses and Calculation of Overall
✓ Equipment Effectiveness
✓ Workplace Improvement
✓ Industry Housekeeping Using 5S
✓ The Visual Factory

Module 4—Quality

✓ The Evaluation of Quality in the Last Fifty Years
✓ What World-Class Quality Means Today
✓ The Seven Basic Quality Tools
✓ Quality Improvement Strategies
✓ Quality Awards
✓ Quality System and ISO 9000
✓ Statistical Quality Control
✓ The Cost of Quality
✓ Defect Tracking and Reporting
✓ Translating Customer Requirements into Quality

Specification

- ✓ Quality Planning and Delivery
- ✓ TQC and the Japanese Perspective
- ✓ TQM and the US Perspective
- ✓ Quality Teams and Teamwork
- ✓ Quality Assessment
- ✓ Customer: the Ultimate Challenge

Module 5—Cost Accounting and Control

- ✓ Basic Concepts
- ✓ Financial vs. Cost Accounting
- ✓ Line vs. Staff Position
- ✓ Variable vs. Fixed Costs
- ✓ Manufacturing Costs: Direct Material, Direct Labor, Overhead
- ✓ Assignment of Overhead Costs: Traditional Methods
- ✓ Activity-Based Costing
- ✓ Cost Information as a Decision-Making Tool
- ✓ Pricing
- ✓ Make vs. Buy
- ✓ Special Orders

Module 6—Reengineering

- ✓ How the Need for Reengineering Developed
- ✓ Business Process Reengineering
- ✓ Concepts and Definitions

- ✓ The Reengineering Process
- ✓ Reengineering, Risks, and Rewards
- ✓ Forward Engineering, Reverse Engineering

Restructuring

- ✓ Software Reengineering
- ✓ How to Evaluate Reengineering Projects
- ✓ Technologies for Reengineering
- ✓ Design Recovery: Business Rules Recovery
- ✓ Future Trends in Reengineering
- ✓ Customer: The Ultimate Challenge

Module 7—Logistics and Production Planning

- ✓ Typical Production Environments Today and Tomorrow
- ✓ Business and Production Planning
- ✓ Production Planning and Activity
- ✓ Bill of Materials
- ✓ Master Production Scheduling
- ✓ Material Requirements Planning: MRP II
- ✓ Manufacturing Policy
- ✓ Production Activity Control
- ✓ A Typical Shop Floor and Routings
- ✓ Work-in-Process and Inventory Turnover
- ✓ Production Types: Capacity Planning
- ✓ Scheduling and Priority Control
- ✓ Lead Time Elements
- ✓ Just-in-Time Manufacturing
- ✓ Supplier Management
- ✓ Finding Waste and Reducing Cycle Time

- ✓ Concurrent Manufacturing Techniques
- ✓ Customer: The Ultimate Challenge

Management Modules:

Module 8—Leadership and Management Fundamentals

- ✓ Five Key Rules of Effective Management
- ✓ Critical Success Factors
- ✓ Self Assessment Including Leader, Coach
- ✓ Communicator/Translator
- ✓ Toolkit to Assist in Strengthening Management Skill
- ✓ A Personal Action Plan for Focus Back on the Job
- ✓ Leadership Skills Explained

Module 9—Planning, Scheduling, and Controlling

- ✓ How to Develop a Business Plan
- ✓ Scheduling, Resources, Risks, and Alternatives
- ✓ Customer, Government, and Competitive Factors
- ✓ Putting the Plan into Action
- ✓ Monitoring, Metrics, and Feedback
- ✓ PERT (Program Evaluation and Review Techniques)
- ✓ How and When to Take Corrective Actions
- ✓ Theory and Practice of Successful Program Management

Module 10—People Management

- ✓ Human Resources Fundamentals
- ✓ What Successful Companies Do for Their Employees
- ✓ Teamwork and Personnel Management

- ✓ Developing Teamwork
- ✓ Time Management
- ✓ Conflict Management

Module 11—Supplier Management

- ✓ Supplier Selection and Qualification
- ✓ Supplier Verification
- ✓ Role of the Supplier as an Integral Part of the Product Team
- ✓ Alternative Suppliers

Module 12—Management Trends

- ✓ ISO 9000
- ✓ Concurrency
- ✓ Accelerating Innovation
- ✓ Technology Trends: Engineering, Manufacturing, Support

Module 13—Cell Biology

- ✓ Cytokine Activity
- ✓ Oxidative Burst
- ✓ Cell Secretion and Cell Contents
- ✓ Procoagulant Activity

Module 14—Pharmacology

- ✓ Drug Discovery
- ✓ Vaccine Production

Module 15—Biochemistry

- ✓ Protein Structure and Function
- ✓ Serine Protease
- ✓ Enzyme Assays

Bioengineering Modules:

Module 16—Introduction to Analysis of Systems with Combined Properties

- ✓ Step Response of a Resistant-Compliant System
- ✓ An Examination of the Step Function
- ✓ An Examination of Step Response Data
- ✓ Use of Step Response Data to Estimate Performance
- ✓ Use of Step Response Data in Identification
- ✓ Concept of the Impulse function

Module 17—System Properties: Resistance

- ✓ The Resistive Property
- ✓ Linear Resistance Analysis
- ✓ Static and Dynamic Resistance
- ✓ Simulation and Analysis Using Static Resistance
- ✓ Simulation and Analysis Using Dynamic Resistance
- ✓ Piecewise Linear Approximations
- ✓ Distributed and Lumped Systems
- ✓ Resistance in Other Systems

Module 18—System Properties: Storage

- ✓ The Property of Storage
- ✓ Systems with Volume Storage
- ✓ Electrical Analog of Compliance
- ✓ Combined Hollow Elastic Elements
- ✓ Cylindrical Elements
- ✓ Simulation of a Thermal System with Combined Properties
- ✓ Analog Study of a Thermal-System Response
- ✓ Storage in a Mechanical System
- ✓ The Electrical Analog of Springiness
- ✓ Dual Representation of the Storage Property
- ✓ Electrical Simulation of Thermal Storage
- ✓ Storage in Thermal Systems

Computing Modules:

Module 19—Computer and Office Systems for Small Businesses

- ✓ Typical Office Systems
- ✓ Networking: Local Area and Wide Area
- ✓ Where Computers Are Required: Where They Are Optional
- ✓ Databases and Data Collection

Module 20—Computer Applications

- ✓ Spreadsheets
- ✓ Engineering

✓ Word Processing
✓ Accounting
✓ Quality
✓ Other Types of Applications

Module 21—Graphical User Interfaces

✓ Types of Interfaces
✓ Reports Formats
✓ When to Use Color
✓ Printing Options
✓ E-mail
✓ Technology trends: Multimedia, Digital Systems, and More
✓ Presentation Format: Effective Presentation Overhead Transparencies

Engineering Design and Analysis Modules:

Module 23—Electronic and Electromechanical Packaging

✓ Compliance and Regulatory Services—EMI/RFI and Safety
✓ Cooling of Electronic Components and Systems
✓ Acoustic-Noise Control: Vibration and Shock Control

Module 24—Product Design and Development

✓ Mechanical-Computer-Aided Engineering
✓ Electronic-Computer-Aided Engineering
✓ Industrial Design/Aesthetics
✓ Manufacturing and Material Selection

Module 25—Machine Design

- ✓ Mechanisms, Mechanical Components, and Structures
- ✓ Finite Element Modeling and Analysis

Module 26—Stress, Vibration, and Shock

- ✓ Finite Element Modeling and Analysis for Stress, Vibration, and Shock
- ✓ Thermal Stress Analysis
- ✓ Modal Analysis and Testing
- ✓ Shock Testing
- ✓ Vibration Measurement

Module 27—Product Design

- ✓ Conceptual Design of Products in Solid Models
- ✓ Finite Element Modeling and Analysis
- ✓ Precision Mechanisms Design and Simulation
- ✓ Laminate Composite Design and Analysis
- ✓ Sheet Metal Parts Design and Analysis
- ✓ Injection-Molded Parts Design and Analysis
- ✓ Drafting, Detail, and Assembly Drawings
- ✓ Fast Prototyping and Cost-Effective Manufacturing

Module 28—Industrial Design/Aesthetics

- ✓ Conceptual Design and Analysis
- ✓ Form and Aesthetics
- ✓ Functionality and Serviceability
- ✓ Ergonomics

✓ Cost-Effective Manufacturing
✓ Sheet Metal Part Design
✓ Injection-Molded Part Design
✓ Material Selection and Fast Prototyping

Module 29—Temperature and Airflow

✓ Chip Level Thermal/Airflow Modeling and Analysis
✓ Board Level Thermal/Airflow Modeling and Analysis
✓ Simulation of Temperature/Airflow in Electronic Enclosures
✓ Junction and Surface Temperature Measurement
✓ Measurement of Airflow and Verification of Distribution Pattern
✓ Component, Subsystem, and System-Level Cooling
✓ Residual, Fatigue, and Thermal-Stress Analysis

The development of this program, to establish the long-term international vision of an interoperable, highly complex, and dynamic information system will be too costly and too risky for even a group of companies to fund.

The technical resources needed to tackle the problem will be diverse. Consortia will be formed as will be shown in following chapter.

CHAPTER 12

Strategic Approach

In this chapter, some modifications will be made to redefine the reason for Blacks present economic condition. Blacks are where they are today because of their limited skills and expertise in scientific—or technological-related subjects, not because they lacked the necessary resources needed to identify them with science- or technology-related products.

As noted in previous chapters Black Africa is the wealthiest in mineral, agricultural, and natural resources in the world, but most of these resources remained obscured and untapped due to limited scientific and technological skills needed to explore and develop these resources into marketable products by Blacks.

In summary, Blacks have abundant resources from which numerous technologies can be developed, but they lack the scientific or technological skills necessary to do so. How to assist Blacks to acquire the requisite skills and knowledge needed to be able to develop technologies from their wealthy resources is an important part of this chapter.

To this point, we have been able to identify the problem and accordingly developed appropriate academic modules needed to tackle the problem. How to integrate the academic modules with the problem identification to achieve the educational objective is the purpose of this chapter.

To ensure that Blacks are united in this effort, universities will be selected from each Black African country. These universities will collaborate with the universities to be identified in United States. Therefore, there will be a joint effort between Black African universities and the United States universities for the first time.

This joint effort is expected to promote unity and teamwork among the Black race to successfully execute this educational program designed to assist Blacks in their quest to explore how to employ their superior and wealthy resources to gain an advantage in competitive world markets.

This program will act as the catalyst for bringing the various Black African institutions and Black African-American institutions to work together as a team. With such a catalyst, true collaborative networks will be possible. Because the development of the program will be a high-risk and long-term venture, the ability to generate the necessary funding needed would be very challenging. The ability to generate matching fund will help minimize the individual risk so that investments can be made in the long-term issues, to accelerate the program. An important consideration is whether or not the Black race alone can successfully fund this program, considering that Blacks have been unable to utilize their talents in the following areas as an edge:

Sports and Entertainment

Table 3, at the beginning of the book, was used to explain the statistical breakdown of the racial composition of the United States league teams, tracks, and boxing. Now we will use the same table to demonstrate that Blacks have misused their talents in pursuing economic security.

	White%	Black%:
Tracks	20%	80%
Football	12%	88%
Basketball	11%	89%
Baseball	30%	70%
Boxing	10%	90%

Table 14: Statistical data showing the domination of Blacks in the following professional sports in the United States.

Based upon the figures on table 14, Blacks dominate sports and tracks. Of 13 percent of the Black population, Blacks are able to dominate tracks with about 88 percent compared to only 20 percent of Whites. In football, Blacks dominate the game with about 88 percent compared to 12 percent of Whites. In baseball, Blacks dominate the game with about 89 percent compared to only 11 percent of Whites. In baseball, Blacks dominate the game with about 70 percent compared to 30 percent of Whites. In boxing, Blacks dominate the sports with about 90 percent compared to 10 percent of Whites. If these figures are compared against that of science/technology, it will appear that Blacks are misusing their talents.

Since the middle of past century, Black entertainers have made significant progress in the following areas of entertainment:

✓ acting
✓ music and singing
✓ dancing
✓ comedy, etc

In spite of this progress, Blacks have not been able to own their own movie production companies or demonstrate any leadership role in this business domain. Unlike major sports events, where Blacks participation is dominant, the same is not true in entertainment.

Even with a moderate record of domination in entertainment-related undertakings, one can still count the number of Blacks that have received the most prestigious award, the Oscar, in this business domain.

The entertainment and sports progress combined together have not created employment opportunities for up to 1 percent of the Black population, considering that Blacks are only 13 percent of the population. This adds to the 11 percent of unemployment problem of Blacks.

All the criticisms notwithstanding, the richest Blacks in the United States belong to entertainment or athletes. Therefore, if one wishes to raise money for any purpose, he or she must first go to this group. Whether or not the Black athletes and entertainers will help in a high-technology project such as this will be another issue to investigate in this book.

During the nineteenth and early-twentieth centuries, eighteen distinguished Black scientists and inventors were particularly involved with science- and technology-related efforts because they knew that Blacks cannot resolve their catastrophic economic tragedy without significant involvement or participation in science and technology.

This group as well as other Blacks did not, however, consider the exploration of the scientific utilization of Black African resources as one of the finest weaponries to build a technology base that could be used to identify Blacks with science- or technology-related products.

These distinguished Blacks included

- ✓ Benjamin Banneker, 1731-1806, Mathematician
- ✓ Andrew J. Beard, 1849—c.1921, Steam Engine Specialist
- ✓ George W. Carver, 1860-1943, Botanist and Agricultural Chemist
- ✓ Dr. Charles R. Drew, 1904-1950, Medical Scientist

- ✓ James Forten Sr., 1766-1842, Ship-guiding Specialist
- ✓ Lloyd A. Hall, 1894-1971, Food Chemist
- ✓ Frederick M. Jones, 1892-1961, Refrigeration Specialist
- ✓ Percy L. Julian, 1899-1931, Organic Chemist
- ✓ Ernest E. Just, 1883-1941, Biological Scientist
- ✓ Lewis H. Latimer, 1848-1928, Electrical Engineer
- ✓ Joseph Lee, 1849-1905(?), Baking Technology Specialist
- ✓ Jan E. Matzeliger, 1852-1889, Shoe Machine Specialist
- ✓ Elijah J. McCoy, 1843-1929, Mechanical Engineer
- ✓ Garrett A. Morgan, 1875-1963, Traffic Signal Specialist
- ✓ Norbert Rillieux, 1806-1894, Sugar Technology Specialist
- ✓ Lewis Temple, 1800-1854, Whaling Technology Specialist
- ✓ Granville T. Woods, 1856-1910, Electromechanical Specialist
- ✓ Louis T. Wright, 1891-1952, Clinical Antibiotic Researcher

In spite of this remarkable science and technology endeavor by these distinguished Black scientists and engineers, Blacks not only failed to capitalize on these already-established scientific and technological foundations, they also entirely ignored the exploration and utilization of any Black African resources, such as the agricultural resources, mineral resources, and natural resources.

These were necessary for superior scientific knowledge and compelling economic power despite powerful evidence that these resources exhibit strong economic benefits.

To persuade these wealthy Black entertainers and athletes to unite as a team and contribute financially to this new technological and scientific crusade, they have to be first convinced that science or technology has the feasibilities to offer enormous economic power to Blacks.

Successful Black entertainers and athletes will have to be convinced that sustainable economic potential of Black African resources can result to a Black-owned firm that could employ more Blacks in a year than both the entertainment and sports industries combined. Furthermore, such an entity could provide an economic return that will be used to educate African and African-American communities on how this technology could be utilized at local level to promote grassroots efforts and improve the general quality of life.

These anticipated economic benefits will be based on medium-sized hi-tech companies. If a medium-sized hi-tech company can create this impressive economic opportunity, imagine the enormous economic opportunities that about five hi-tech companies the size of Boeing or Lockheed Martin or IBM can create for Black communities.

While employees of these Black hi-tech companies may not be at the top of the economic ladder, their earnings will certainly eliminate them from poverty status and enable them to help other Blacks as well, thereby establishing economic security among Blacks and other ethnic minorities.

The goal of this book is to ascertain that Blacks recognize all opportunities available to him and wisely capitalize on such opportunities to tap into the equation of science or technology that is very much needed to identify Blacks with science- or technology-related products as they stride to gather forceful momentum needed to accelerate them into the twenty-first century with superior economic status.

History has shown that the American Indians and others who occupied the United States before it was discovered were unable to satisfy the manual labor requirements of the arriving Europeans. This forced these Europeans to look for a different group of people to meet their labor requirements. Subsequently, the Europeans discovered that Blacks exhibited the superior skills required.

These superior skills present another opportunity given to Blacks, which they failed to use to develop the strong confidence necessary to control and commend most of the resources in the United States today. These skills, in conjunction with abundant African wealth such as the Black African resources discussed previously, should have given Blacks the command of economic power. In spite of these superior characteristics, the creditability, integrity, and prestige of Blacks has been significantly damaged and suppressed by Whites.

The most troubling scenario with this concept is that Blacks have been deceived beyond reasonable doubt that they are not as bright as other races and should therefore focus their entire effort to specific areas of life.

Although Blacks have failed to utilize these opportunities to improve their economic and academic status, this trend does not have to occur. Today, Blacks must advocate, through all means necessary, to their youths that appropriate skills in science, engineering, communication, and basic manufacturing can play critical roles in assisting Blacks in their struggles to gain superior economic status. The skills required are shown below.

Science and Technology Goals and Missions:

1. Strengthening youths' commitment to remain in school and continue the study of science, mathematics, and engineering, emphasizing science and engineering as alternative skills in meeting the challenges imposed by the fast changing business world.
2. Illustrating the importance of science, mathematics, and communication skills in everyday lives.
3. Exposing the youths to a broad range of participatory activities in science, mathematics, and engineering, including research and interactions with engineers and scientists.
4. Offering the youths information and guidance in the career-exploration process including the academic preparation necessary for a variety of professions in science and engineering.
5. Emphasizing *science* and *engineering* as sources of hope and potential economic vitality.
6. Emphasizing science and technology as an alternative approach in narrowing the wide margin of the existing economic gap between Blacks and Whites, which can bring about social unity among all races.

Communication Goals and Missions:

I. Interface with people of diverse educational and cultural backgrounds:

    ✓ Recognize that every one can make a contribution
    ✓ Prejudice impedes problem solving and impairs productivity

II. Communicate in written and spoken English:

    ✓ Good ideas must be presented effectively
    ✓ Individuals must be able to articulate their points of view

III. Manage time, prioritize, and think strategically:

    ✓ In an empowered work environment, appropriate skills are necessary

IV. Translate theory into practice:

    ✓ Solving problems requires conceptual knowledge on a specific problem
    ✓ Learning how to frame questions and derive solutions is skill developed by practice

V. Use science and math, including statistics:

    ✓ To analyze problems and continuously improve productivity, scientific and engineering knowledge are critical elements

VI. Employ ever-evolving computer-driven tools:

✓ Automation affects design, production, sales, accounting, and administration
✓ Small businesses need to know how to use the latest tools to maximize efficiency

VII. Sustainable environmental development:

✓ Production planning and design must address the impact of economic, cultural, environmental, and technological factors on sustaining the total ecosystem

Basic Manufacturing Goals and Missions:

I. Manufacturing
II.
✓ Integrated manufacturing
✓ Intelligent processing
✓ System management technologies
✓ Product design
✓ Product development
✓ Manufacturing
III. Materials
IV.
✓ Materials development linked to performance
✓ Synthesis and processing of materials
✓ Ability to deploy from a wide group of materials such as metals and alloys, ceramics, polymers, and composites

- ✓ Exploitation of new materials for electronic, magnetic, and photonic applications
- ✓ Thin film processing

Information and Service

- ✓ Development of software engineering
- ✓ Storage products
- ✓ Display
- ✓ Computer simulation and modeling

IV. Integration of Technology

- ✓ Technology transfer
- ✓ Miniaturization of products and device
- ✓ Systems automation
- ✓ Micromechanics
- ✓ Systems control (electromechanical)
- ✓ Multimedia technologies

V. Environment

- ✓ Efficient use of energy and materials
- ✓ Waste minimization
- ✓ Pollution prevention
- ✓ Alternative materials and processes
- ✓ Recovery and recycling
- ✓ Sensors to monitor materials

The progress of Blacks in entertainment and athletics has not been very significant, considering that Blacks have not dominated these business domains or commanded any sort of academic or economic superiority or prestige in these business areas.

However, when the progress of Blacks in this area is compared to their progress in other areas, it could be said that their progress is a bit encouraging. Unfortunately, however, Blacks' continued failure to point their youth in the direction of science or technology is indirectly misleading the Black youth.

Black youth believe Blacks are only good in sports—or entertainment-related activity. As a result, they continued to pursue carriers as entertainers or athletes, believing that these are the only things Blacks are good at.

Consequently this creates problems as Black youngsters lack much motivation to pursue careers in mathematics, computer science, science, and engineering, see table 1. An assumption can be made that Black entertainers or athletes are indeed not doing much to help the economic status of Blacks. Their only belief is individual achievement, as are many other Blacks. This very problem is worsened by the news media. Almost every report about the Black race is negative.

On the television, Blacks are portrayed only as entertainers, as a result, young Blacks think that is the only thing Blacks can aspire to. Consequently, our Black youngsters ignore pursing carriers in science or engineering. No Black child watching the television has ever seen Black engineers or scientists on the television. If the problem stops here, it may not be very damaging as the present situation. But the problem goes beyond that.

CHAPTER 13

Superior Economic Status

In this chapter, certain key elements need to be reemphasized or repeated to justify the concept that technology is the key to economic security and the only logical approach for Blacks to enjoy improved economic power. It is, therefore, worth repeating that key advancements made by Blacks to this point include

- ✓ the expulsion of Europeans from Black Africa, which resulted to the independence of many Black African countries. Dr. Nnamdi Azikiwe successfully led this crusade.
- ✓ the restoration of social equality and justice to the Black African-American. Dr. King successfully led this crusade.

These two major achievements were possible due to the fact that Blacks, in both cases, performed as a team and fought for a common purpose.

The most challenging aspect of the process will be how to generate the funds needed to successfully achieve this goal.

Considering the enormous wealth of various Black individuals, generating the necessary funds needed to successfully create hi-tech companies of reasonable sizes should not be very difficult.

However, considering how difficult it is for Blacks entertainers and athletes to help other Blacks, such as Black scientists or engineers, the successful accomplishment of this effort will be very difficult and challenging.

The most difficult and interesting aspect of this chapter is to successfully develop a methodology that will unite Blacks from both sides of the equation to work together in this effort. A prominent influential Black with access to Black celebrities will be sought. This prominent Black will work hand in hand with knowledgeable Black scientists or engineers.

This joint effort will be aimed toward generating the necessary funds needed to accomplish this goal. Persuasive presentation will be made by the scientists or engineers. Arrangements will be made to get some of the celebrities together at a time. Why it is necessary to make the move now will be explained.

The enormous economic benefits associated with technology development will be addressed to the satisfaction of these celebrities. Lastly, they will be assured that their investment is wise and will bring maximum profits in return.

Their inputs to the technology development will also be assessed for utilization in the development phase of the project. African technologies or resources, certainly most appropriate for this effort, will be emphasized over and over because such technology carries with it the following economic benefits and opportunities:

1. Improved understanding of obscure African technologies by objectively exploring their superior scientific characteristics for effective utilization in world market.
2. Established analytical and design fundamentals necessary to identify American products that can be very profitable to African economy by applying basic engineering fundamentals—such as fluid dynamics, stress, aerodynamics, and thermal analysis—for proper verification and identification of model deficiencies, design adequacy, proper performance, and performance anomalies to promote exportation opportunities to Black Africa as well.

CONCLUSION

I believe this is a very important book. Based upon my knowledge and experience, I have not seen a book written to sincerely define the cause of Black poverty from the science and technology perspective. If Blacks plan to enjoy the economic power they have been deprived of, for hundreds of years, they must change their current strategy and focus extensively on the approach advocated in this book. Science and technology that will result to a technology-related product will give Blacks the recognition they very much deserve. It will also fashion them with the needed economic power. I hope this book will make a difference in your life. Thank you for your support.

Curriculum Vitae

Raymond L. Chukwu
Black Technologies Advancement
1190 Saratoga Ave. #150
San Jose, CA 95129
408 244 7852 btagrant@comcast.net

Raymond has been described as an American success story. Here is why. He was born in California to an American mother and a Nigerian father. His father, saddened and disheartened by the death of his, mother took him to Nigeria when he was two years old to allow him to grow up with paternal relatives. His grandmother was delighted to finally have a chance to become acquainted with her grandson she has never seen and delightedly agreed to take care of Raymond while his father went to work in the city.

Tragedy once again struck shortly after his arrival to Nigeria, however. It was on his father's way to his new position few days later that he was killed in a truck accident. As a result of his sudden death, he had had no chance to tell his relatives, particularly Raymond's grandmother, anything about Raymond's

---

side of the family other than the simple fact that he had been born in Sonoma, California.

Even this fact was kept from Raymond when he was growing up by his Nigerian relatives, including his grandmother, until one day, it slipped out in conversation. From that day on, Raymond knew he would come back to the United States to see the place he was born.

His grandmother who had raised him so lovingly taught him the following important life reasons at a very young age before her dead when Raymond was twelve years old:

- ✓ strength
- ✓ wisdom
- ✓ uncommon diligence

After her death, other relatives took over her work of taking care of Raymond while he attended private Nigerian schools. These relatives and others had told Raymond that it would be impossible for him to return to the United States and accomplish any of his goals. Much was to occur before Raymond was finally able to return to the United States to fulfill his lifelong dream of becoming an aerospace engineer.

Since his return home to California, Raymond has succeeded with the following:

- ✓ 2006 Republican nomination for US Congress
- ✓ 2004 Republican nomination for US Congress

- ✓ 2001 being a finalist competitor for the post of US ambassador to Nigeria
- ✓ Stanford University/Lockheed Graduate Engineering Program
- ✓ Lockheed Graduate Engineering Program
- ✓ BS Aerospace Engineering
- ✓ Being author of four books
- ✓ Being the founder and CEO of Black Technologies Advancement (BTA)
- ✓ Being the founder and CEO of Bio Chemical Technology (BCT) Aerospace
- ✓ Being the founder and CEO of BTA Foreign Student Exchange Scholarship program
- ✓ Given that relatively little scientific data are available today to convincingly demonstrate whether or not a particular treatment with native African Herbs is safe, efficacious, beneficial, helpful, or leads to a positive outcome (e.g. produces a regression in the size of a tumor, prolongs or improves quality of life, reduces or eliminates adverse symptoms of a toxic treatments), his commitment has been to generate such scientific data and assess their impact on the pharmaceutical and biochemical industry.

These accomplishments are targeted at giving something in return. Creating social and strong economic opportunities between the United States and sub-Saharan African countries with Nigeria as the model will certainly change the dynamics of global marketplace.

In parallel with this viewpoint in 2000, when George W. Bush became the United States President, his wife Debbie suggested that he seek employment with his administration. At first Raymond did not give this much consideration.

However, given his strong commitment to Black's economic strength through technological sophistications, he figured that seeking the appointment of the United States ambassador to the largest Black African nation (Nigeria) would be right on target. Therefore he decided to seek the post of United States ambassador to Nigeria.

That seemed laughable at first, considering both his economic and political influence. The task seemed impossible, but he has nothing to lose but has everything to gain by trying.

In his book My American Dream from an Orphan to an American Success Story, you will read step-by-step on how he went from Raymond Who to actually becoming a very powerful finalist competitor for the appointment of the United States ambassador to Nigeria, which was only interrupted by the event of September 11, 2001.

You will also see how he tactically, through effective correspondence, build and develop strong political relationship and friendship with the most influential and powerful United States senators and the vice president, which created the advantage for him to become the finalist competitor for the appointment of the US ambassador to Nigeria.

Furthermore, you will read how this astonishing performance created an instantaneous opportunity for him to be considered as the most credible candidate in his district to run and receive the nomination of his party for United States Congress in the Fifteenth US Congressional District, California.

Publications:

He has written numerous articles in Nigerian newspapers advocating the need to explore abundant agricultural and mineral resources of the country as an alternative exporting product in place of oil. He completed a book on wind power as an alternative energy use in African universities.

Independent research study on scientific utilization of black soap derived from native African herbs (NAH) as cost-effective countermeasure against nuclear—and space-radiation-related sickness. Published by Lockheed Martin in Sunnyvale, California, in 1986.

He has done research study on developing pharmaceuticals from native African herbs (NAH) as cost-effective alternative medicine. Published by Bio Chemical Technology (BCT) Aerospace, San Jose, California, in 1992 and the East Bay Monitor, Oakland, California, in May 1991.

BLACK TECHNOLOGIES
ADVANCEMENT

Black Technologies Advancement (BTA) was established in 1992 as a 501(c)(3) nonprofit research and educational organization. It was created to specialize in advance scientific research aimed toward improving the scientific knowledge on developing new pharmaceuticals from natural products, especially from native Black African herbs (NBAH). Developing new methods to scientifically analyze, for the first time, data concerning native Black African herbs (NBAH) for treatment against serious diseases of concern to Black Africa and the United States is a key core competency within the NBAH Research Division at BTA.

The research staff at BTA use advanced knowledge and experience in natural products, analytical and computational chemistry, biology, biochemistry, and medicine to conduct its biomedical research on developing new pharmaceuticals from natural products such as native Black African herbs (NBAH). In addition, BTA is in collaboration with Bio Chemical Technology (BCT) in San Jose, California, in its effort to successfully develop new pharmaceuticals from NBAH. Because of BCT's impressive track record in biomedical and scientific research over the years, BCT is well respected and recognized as a leader in elementary analysis of agents derived from natural products. Furthermore, BCT is in collaborative effort with Lawrence Livermore National Laboratory (LLNL), Sandia National Laboratories (SNL), and San Jose State University (SJSU) in most of its scientific research. These expertise and the commendable business relationship with key government laboratories and universities are expected to be very advantageous to the proposed effort. Internationally, BTA is in collaboration with various Black African universities, Igbo Traditional Medicine Center (Nigeria), and the Medicinal Plants Project/CISED (Dakar, Senegal).

BTA will work closely with both the domestic and international organizations, in an atmosphere of mutual respect and empathy, to successfully prove the concept of the proposed effort and ensure that the resulting technology addresses the economic need of Black Africans and the United States.

Given that drugs derived from plants is expected to crowd the pharmaceutical market in the immediate future, it was necessary to assess whether any of these benefits could be attributed to the drugs derived from the native Black African special-illness-healing herbs. It was clear from our assessment that indeed the pharmaceutical properties of native Black African herbs (NBAH) have remained obscured to the scientific and biomedical communities. There is, therefore, a scientific and economic justification to develop new methods to evaluate whether the pharmaceutical properties of the native Black African special-illness-healing herbs will have any measurable impact on the world pharmaceutical market. Through successful identification and initial screening of compounds from native Black African herbs (NBAH), we could derive the following scientific and economic benefits:

- ✓ scientific data that could lead to bioassay-guided NBAH chemistry
- ✓ advancement of NBAH pharmacologically value
- ✓ promote pharmaceuticals from NBAH as biotechnology cornerstone

The scientific advancements and technological feasibilities resulting from this new methods is expected to provide the following technological and economic advantages to the United States and Black Africa:

- ✓ biotechnology-related investment opportunities
- ✓ profitable business relationships
- ✓ wide utilization of NBAH as an alternative medicine
- ✓ new scientific data to advance medicinal and medical knowledge
- ✓ identification of Blacks with science- and technology-related products
- ✓ identification of Black Africa as a potential component of the world economy

The objective of Black Technologies Advancement in its effort is to gather and analyze the pharmaceutical properties and bioactive agents of native Black African herbs (NBAH) as new alternative treatments. Through performing the identification and initial screening of leads with native Black African special-illness-healing herbs, we can analyze a set of

NBAH derivatives and their biological or medicinal properties as inputs to our model prototype program to generate outputs of various combinations of these derivatives. By determining which combinations that combined successfully, we can discover which NBAH combination or combinations exhibit activity against specific disease of interest. Through programming and coding these mathematical equations, we can further verify and predict the biological activity of each NBAH against a specific disease of interest.

Index

A

B

H

Hall, Lloyd A., 121
health care treatment, 13, 90
  benefits, 91
  Blacks' medical care, 15
  good health and superior medical
    care, 17
  health care budget, 70
  hospital treatment, 70
  poor health care, 13
hi-tech companies
  Boeing, 122
  IBM, 122
  Lockheed Martin, 122
highly skilled labor force, 24
high blood pressure, 74
  attributed to skin color, 75

I

independence, 49
industrialized nations, 70
Information and Service, 127
integrity, 29

J

Jones, Frederick M., 121
Julian, Percy L., 121
Just, Ernest E., 121

K

King, Martin Luther Jr., 19, 99, 129

L

Latimer, Lewis H., 121
Lee, Joseph , 121
limited scientific and technological
    skills, 13

M

market, 27
  market base, 29
Matzeliger, Jan E., 121
McCoy, Elijah J., 121
medical technologies, 15
medicinal procedures, 15
Middle America, 28
misdiagnoses, 77
modern medicine, 49
Morgan, Garrett A., 121

N

National Center for Health
    Statistics, 84
native Black Africans, 18. *See* also
    Blacks (race)
native Black African herbs, 18, 46,
    58, 60, 81
native sub-Saharan African
    herbs and plants, 18
NBAH. *See* native Black African
    herbs
NBAH-Butter, 68

www.ingramcontent.com/pod-product-compliance
Lightning Source LLC
Chambersburg PA
CBHW051245050326
40689CB00007B/1077